Law, Governance and Technology Series

Volume 64

The *Law, Governance and Technology Series* is intended to attract manuscripts arising from an interdisciplinary approach in law, artificial intelligence and information technologies. The idea is to bridge the gap between research in IT law and IT-applications for lawyers developing a unifying techno-legal perspective. The series will welcome proposals that have a fairly specific focus on problems or projects that will lead to innovative research charting the course for new interdisciplinary developments in law, legal theory, and law and society research as well as in computer technologies, artificial intelligence and cognitive sciences. In broad strokes, manuscripts for this series may be mainly located in the fields of the Internet law (data protection, intellectual property, Internet rights, etc.), Computational models of the legal contents and legal reasoning, Legal Information Retrieval, Electronic Data Discovery, Collaborative Tools (e.g. Online Dispute Resolution platforms), Metadata and XML Technologies (for Semantic Web Services), Technologies in Courtrooms and Judicial Offices (E-Court), Technologies for Governments and Administrations (E-Government), Legal Multimedia, and Legal Electronic Institutions (Multi-Agent Systems and Artificial Societies).

Larisa V. Sannikova
Editor

Digital Technologies and Distributed Registries for Sustainable Development

Legal Challenges

Springer

Editor
Larisa V. Sannikova
State Academic University of Humanities
Moscow, Russia

ISSN 2352-1902 ISSN 2352-1910 (electronic)
Law, Governance and Technology Series
ISBN 978-3-031-51066-3 ISBN 978-3-031-51067-0 (eBook)
https://doi.org/10.1007/978-3-031-51067-0

This Springer imprint is published by the registered company Springer Nature Switzerland AG
The registered company address is: Gewerbestrasse 11, 6330 Cham, Switzerland

Preface

The global challenges facing humanity here in the beginning of the twenty-first century are being defined to a great extent by our willingness and determination to achieve the Sustainable Development Goals set forth by the United Nations in 2015. There are 17 SDGs to be achieved by 2030: (1) No poverty; (2) Zero hunger; (3) Good health and wellbeing; (4) Quality education; (5) Gender equality; (6) Clean water and sanitation; (7) Affordable and clean energy; (8) Decent work and economic growth; (9) Industry, innovation, and infrastructure; (10) Reduced inequalities; (11) Sustainable cities and communities; (12) Responsible consumption and production; (13) Climate action; (14) Life below water; (15) Life on land; (16) Peace, justice, and strong institutions; and (17) Partnership for the goals.

The application of digital technology contributes significantly to the achievement of these goals. According to the Global Enabling Sustainability Initiative (GeSI) and Deloitte report #DigitalWithPurpose-Delivering a Smarter2030: "these technologies, if deployed with positive societal impact in mind, will help accelerate progress toward the SDGs by 22 percent and mitigate downward trends by 23 percent on average."[1]

The primary obstacle to the introduction and use of digital technology is legal uncertainty. Legislators in technologically advanced countries are cautious in their attempts to legalize and promote the use of potentially intrusive digital technologies, seeking to minimize the risks associated with the public's lack of knowledge, and in some cases fear, of such digital novelties.

This book attempts to identify and explore the legal issues raised by the use of digital technologies to achieve the SDGs. The authors focus primarily on Distributed Ledger Technologies and Artificial Intelligence, both of which have been in widespread use for some time. These technologies are transforming different areas of public life, such as finance, industry, energy, education, justice, etc. The authors are

[1] GeSi, 2019. https://gesi.org/research/download/36

not limited to researching the positive effects of digitalization on the achievement of the SDGs, they carefully analyze the negative ones as well.

However, at the moment the issue of the challenges of digitalization in achieving the SDGs has been put aside. The main obstacle to the world's path to sustainable development has been the aggressive actions of politicians who have brought the world to the brink of nuclear war. In these circumstances, the role of the academic community is more important than ever. During the Cold War, the Club of Rome was able to bring intellectuals together to solve global problems. This example should inspire scientists from different countries to come together to find joint solutions for peace and sustainable development.

This book is the result of such an inspiring collaboration of scientists from different countries and continents. The editor would like to express his sincere thanks and appreciation to all the authors of the book for their cooperation. The editor is grateful to Prof. Dr. Julia Kharitonova for her help in communicating with the authors. The editor also thanks Prof. Dr. Bernd Holznagel for his support in publishing the book.

Moscow, Russia Larisa Sannikova

Acknowledgment

This research was supported by the Russian Foundation for Basic Research via grant No. 18-29-16145 MK "The mechanism of legal regulation of relations using the Distributed Ledger Technology."

Contents

Part I Artificial Intelligence and Sustainable Development

An Assessment of the Role of Artificial Intelligence on Sustainable Development Goals 3
Prashant Tripathi and Priti Saxena

Sustainable Growth and the Role of Artificial Intelligence in Improving the Circular Economy 25
Carlo Amatucci and Giovanni Mollo

Part II Blockchain for Sustainable Development: The Potential of Tokenization

Token, Tokenization and Sustainable Development 45
Raffaele Lener and Salvatore Luciano Furnari

Tokenization of the Creative Industries: The Intersection Between Emerging Technologies and Sustainability 59
Yulia Kharitonova

Part III The Green Digital Transformation: Legal Issues

The Importance of a Global Legal Framework and Digital Technologies in Combating Climate Change 79
Cocou Marius Mensah and Tjaša Ivanc

Green Finance in the EU and Russia: Legal Frameworks and Opportunities of Digitalisation 97
Larisa V. Sannikova

Part IV Human Rights and Information Communication Technologies

Fast Internet as a Prerequisite for Sustainable and Resilient Development: Network Expansion Measures and Extension of User Rights 113
Bernd Holznagel

The Right to Education in a Digital Era 135
Suzana Kraljić

Part V Challenges in Achieving SDG 16

Corporations and the Global Backlash Over Privacy: Too Big to Regulate? 157
Eduardo Saad-Diniz and Matheus Puppe

E-Justice in Russia 167
Vasily Laptev

Part I
Artificial Intelligence and Sustainable Development

An Assessment of the Role of Artificial Intelligence on Sustainable Development Goals

Prashant Tripathi and Priti Saxena

Abstract Association of Artificial Intelligence (AI) with the Sustainable Development will make India a better tomorrow and help industries of almost every sector to construct a better planet. Much attention is currently required for the steadily emerging concept of Artificial Intelligence and its wide impact upon different sectors of the society. Also, need for evaluation of its impact upon sustainable development is the need of the hour because they are both positive and negative impacts. The categorization of all the 17 Sustainable Development Goals (SDGs) was on the line of economic, social and environmental goals by United States in 2015. SDGs provide for the innovative and practical blueprint for understanding the positive and negative effects of AI. In the medium and long term, AI is projected to have an impact on global productivity, equality and inclusion, environmental outcomes, and a variety of other areas. AI's reported potential consequences on sustainable development are both positive and negative. They're also having an impact on global sustainability trends. As the AI revolution affects our world, it may announce a utopian future in which humans and machines coexist peacefully, or it may herald a nightmarish future filled with conflict, poverty, and pain. More immediately, would AI help us achieve the Sustainable Development Goals (SDGs) of the United Nations (UN) or push us further down the road to more economic Uncertainty, environmental collapse, and societal upheaval turning out as the possible outcomes which is to be addressed as the need of the hour. The purpose of this article is also to answer these issues by examining the effects of AI-SDGs relationship and to look into the conceptual framework presented here which aids in the investigation as to which SDGs AI may be useful in attaining and which goals may be compromised as a result of increased AI use.

Keywords Sustainable development goals · Artificial intelligence

P. Tripathi · P. Saxena (✉)
Babasaheb Bhimrao Ambedkar University, (A Central University), Lucknow, India

L. V. Sannikova (ed.), *Digital Technologies and Distributed Registries for Sustainable Development*, Law, Governance and Technology Series 64,
https://doi.org/10.1007/978-3-031-51067-0_1

1 Introduction

Technologies are expected to consolidate as a societal and economic lever for transformation at global scale. Artificial Intelligence (AI) has currently reached a sufficient level of maturity both as an umbrella scientific discipline and as a technology, having spread from laboratories to the entire society (Palomares et al. 2021).[1] Artificial Intelligence (AI) is rapidly expanding into new domains such as health, poverty, commerce, agriculture, education, disaster management, etc. and their respective government policies. Machine intelligence and robotics with deep learning capabilities have had significant disruptive and enabling effects on almost all the areas of the society. It is widely asserted, Artificial Intelligence (AI) has been enormously promising and has seemingly proven, that it has the ability to change the world for its betterment. Despite widespread enthusiasm, many people remain skeptical, both of the evidence purporting to illustrate AI's efficacy and of the numerous potential applications leading to the negative impacts of the Artificial Intelligence. As a result, AI ethics has grown into a thriving field of research, but it is still in its fancy, and there is not much agreement on what it involves or how it should be pursued, as with other new areas (Lideniro 1991).[2]

The United Nations (UN) established 17 Sustainable Development Goals (SDGs) in 2015, which must be met by 2030.These include environmental, social, and economic/financial objectives. The SDGs provide a good framework for examining and categorizing the possible benefits and downsides of AI as technology becomes more common in modern cultures (Population Matters, 2021).[3] The SDGs are utilized as a framework for examining the entire implications of AI on concerns of sustainability in the broadest sense, and one of the article's primary contributions is to look into the fact that how SDGs acts as a technique for ethical evaluation of AI.

2 Concept and Historical Background of Sustainable Development Goals

Sustainable Development is described as a way of development that satisfies current wants and aspirations without jeopardizing future generations' ability to fulfil their own needs and desires (International Institute for Sustainable Development 2020).[4] Sustainable Development is one of the most widely debated subjects in the world

[1] Palomares et al. (2021), pp. 6497–6527.

[2] Alegre Lideniro (1991), pp. 1306–1309.

[3] Population Matters. 2021. "Population and the Sustainable Development Goals". https://populationmatters.org/sdgs?gclid=Cj0KCQjwraqHBhDsARIsAKuGZeH57nTpQ2Eiglr1Koj9fol8HEfcYC1vcLZIKbtvkhBbODb8vFmiSrIaAnesEALw_wcB (last updated July 12, 2021).

[4] International Institute for Sustainable Development (2020). "Sustainable Development | International Institute for Sustainable Development". Last updated July 12, 2021. https://www.iisd.org/about-iisd/sustainable-development.

today, since it is the best alternative to development techniques that include extensive deforestation and the loss of natural resources.

The aim of the SDGs is basically guiding the world for looking forward to a more safe, sound and effectively peaceful environment with more sustainable consumption and production patterns, while also enshrining a global commitment "to leave no-one behind": not in the delivery of services nor when it comes to engaging people in decision making. For the development community, 2015 was a "once-in-a-generation year," as global commitments were made that will have a significant impact on people's and our planet's futures (United Nations 2005).[5] The agreement on the 2030 sustainable development agenda "Transforming our World," which was ratified by all UN member states at a Special Summit in September 2015 in New York, was the most important among them. This agenda for peace, prosperity, planet, people and partnerships will influence national policies, related budget allocations and will continue for many years to come. With 17 Sustainable Development Goals (SDGs) and 169 targets, the 2030 Agenda for Global Action builds on the legacy of the eight Millennium Development Goals (MDGs), which focused on poverty and hunger, health, education, gender equality, the environment, and global partnerships. Unlike the MDGs, which were aimed at poor nations, the SDGs are universal and apply to all Member States (The World Bank 2015). [6] They are also far more ambitious than the Millennium Development Goals, as they address not only the social, but also the economic, environmental, and political dimensions of sustainable development, with goals on inequalities reduction, infrastructure, energy, peaceful societies, and other new areas (Sachs 2012).[7] The United Nations Conference on Sustainable Development in Rio de Janeiro in 2015 gave birth to the Sustainable Development Goals (SDGs). The purpose was to create a set of universal goals that addressed the world's serious environmental, political, and economic concerns. The Sustainable Development Goals (SDGs) replace the Millennium Development Goals (MDGs), which began a global effort in 2000 to address indignity of poverty (United Nation Development Program 2020).[8]

For 15 years, the MDGs have driven progress in several key areas, including reducing income poverty, increasing maternal health, and providing much-needed access to water and sanitation. They also inspired countries to invest in future generations by launching a global campaign for free elementary education. The MDGs' legacy and accomplishments equip us with invaluable insights and

---

[5]United Nations (2005). "Department of Economic and Social Affairs| Sustainable Development." Last updated July 12, 2021 https://sdgs.un.org/goals.

[6]The World Bank (2015). "Millennium Development Goals," accessed July 12, 2021, https://www5.worldbank.org/mdgs/.

[7]Sachs (2012), pp. 2206–2211.

[8]United Nation Development Program (2020). "Sustainable Development Goals." https://www.africa.undp.org/content/rba/en/home/sustainable-developmentgoals/background.html#:~:text=The%20Sustainable%20Development%20Goals%20(SDGs,economic%20challenges%20facing%20our%20world.

experience as we begin work on the new goals (UN Department of Public Information 2015).[9]

However, The SDGs are a bold promise to finish what we started and address some of the world's most pressing issues. All 17 goals are interconnected, so success in one has an impact on the others. Dealing with the threat of climate change has an influence on how we manage our finite natural resources, attaining gender equality or improved health aids in poverty eradication, and encouraging peace and inclusive communities reduces disparities and aids economic growth. In a nutshell, this is the best opportunity we have to better the lives of future generations. The Sustainable Development Goals (SDGs) are unique in that they address challenges that impact everyone. They reaffirm our international resolve to eradicate poverty in all of its forms, forever. They are determined to ensure that no one is left behind (Vinuesa et al. 2020.).[10] More significantly, they enlist our participation in the creation of a more sustainable, safer, and wealthy world for all of mankind. The SDGs are intended to carry on the work of the MDGs in improving the world by making bolder decisions and strategies in response to the issues that the world faces today (Sætra 2021).[11]

As part of the 2030 Agenda for Sustainable Development, the 'Sustainable Development Goals' (SDGs) are to be implemented internationally for the good of the planet. There are 17 Sustainable Development Goals (SDGs) aimed at reducing poverty, hunger, and inequality, as well as combating climate change and other issues.

3 Notion of Artificial Intellegence

This article is based in part on a wide definition of AI, which implies that AI encompasses a diverse set of technology and applications. ***"Any software technique with at least one of the following features was termed AI, i.e., perception—including audio, visual, textual, and tactile (e.g., face recognition), decision-making (e.g., medical diagnosis systems), prediction (e.g., weather forecast), automatic knowledge extraction and pattern recognition from data (e.g., discovery of fake news circles in social media), interactive communication and logical reasoning*** (Negnevitsky 2005).[12]" According to Nils J. Nilsson: AI is that activity devoted to making machines intelligent, and intelligence is that quality which enables an entity to function appropriately and with foresight in its environment

[9]UN Department of Public Information (2015). "We Can End Poverty, 2015" https://www.ndi.org/sites/default/files/Handout 7 - Millennium Development Goals.pdf.

[10]Vinuesa et al. (2020), pp. 1–10.

[11]Sætra (2021), pp. 1–19.

[12]Negnevitsky (2005).

(Jadhav 2019).[13] AI is here viewed in this context rather than as a stand-alone technology (neutral and decoupled). Rather than doing an isolationist examination of AI, this requires viewing it as part of a sociotechnical system that includes numerous structural, economic, and political systems (Spector 2006).[14] The computational infrastructure and non-universal access to ever massive data sets are inextricably intertwined in modern AI. Non-universal access to ever massive data sets, as well as the computing infrastructure required to make use of them, are fundamental to modern AI. Basically, AI is based on the combination of Machine Learning and Deep learning which can be considered as two pillar of AI. One side, Machine Learning is a type of data mining technique that examines the large pre-existing data and extract new valuable information from that database and other side, Deep learning is a subset of machine learning and is called deep learning because it makes use of deep neural networks. The machine uses different layers to learn from the data. The depth of the model is represented by the number of layers in the model.

AI, through a broad spectrum of branches and applications, will impact corporate and business integrity, corporate governance, distribution of financial products and services, intellectual property rights, privacy and data protection, employment, civil, criminal and contractual liability and a significant number of other legal fields. In the area of the manufacturing and industrial process, AI Entities i.e. Robots are replacing the labor forces and performing a heavy activity with more proficiency. For example, Amazon Online Shopping Company has introduced new robotic equipped, Corton Wrap Machine in its warehouses with the aim of creating a cheaper and more efficient supply chain. This Robotic Machine can box 500–600 boxes per hour which are about 4-5 times more than manual human packing (Shield 2019).[15] Similarly, in the field of healthcare, AI is playing a very vital role in order to analyze the complex and large amount of data that was undertaken previously to aid the discovery of various drugs. Recently Peter Rogan, Professor at the University of Western Ontario, has confirmed that AI can be used to predict breast cancer (Borukhovich 2015)[16] which would analyze improvement while using the drug Paclitaxel.[17] This article lays forth a framework for determining which SDGs, AI could help us in achieving and which objectives might be jeopardized by expanded AI use, either directly or indirectly.

[13] Jadhav (2019).

[14] Spector (2006), pp. 1251–1253.

[15] Shield, Nicholas (2019). "Amazon Is Introducing New Warehouse Robots." https://www.businessinsider.com/amazon-introduces-new-warehouse-robots-2019-5?IR=T.

[16] Borukhovich, Eugene. (2015). "How Will Artificial Intelligence Change Healthcare?" https://www.weforum.org/agenda/2015/10/how-will-artificial-intelligence-change-healthcare (last updated July 15, 2021).

[17] Paclitaxel is a chemotherapy drug used to treat ovarian, breast and non-small cell lung cancer.

4 Linkage of AI with SDGs and Its Impact: A Contemplation

AI and the SDGs have been linked in the past. As per the relevant data in extant, AI has the potential to achieve its objectives across the SDGs, mostly through technical advancement. However, the development of AI may have a detrimental influence on few SDGs. for that reason we have divided the SDGs into three categories, based on the three pillars of sustainable development, namely social development, economical development, and environmental development. Sustainable-AI development is an inclusive framework. It is, nevertheless, distinct from previous frameworks in that it has the capacity to emphasize certain features of artificial intelligence and digitalization (Ravallion 2016).[18] This shift in viewpoint provides particular insights that may be applied while debating the future of artificial intelligence. The dialogue on sustainable development has reached a very high degree of inclusiveness. It is applicable to all countries and areas of the world. It tackles economic development challenges while also emphasizing social and environmental concerns. It emphasizes human rights while also being attentive to problems at the group and country levels. It is aimed at both public and private actors and is inclusive of civil society. What's more important is, sustainable development operates at the micro and macro levels. For some goods and uses, sustainability might be a concern. However, it may also be a benchmark for assessing total environmental or social justice consequences. In contrast, the focus of the AI discourse tends to concentrate on the consequences of individual applications, specifically, their performance in terms of justice, accountability, and openness (Natural Capital Partners 2021).[19] With the exception of the future of employment, the effects of AI on society as a whole are hardly considered. As a framework, sustainable AI development has the potential to change this (Foer 2017).[20] One of the remarkable features of sustainable development is its focus on governance and implementation. A sustainable AI approach not only highlights aims and boundaries, but also necessitates the issue of how they will be fulfilled. From this vantage point, it is not enough to talk about ethics and human rights; rather, concrete measures must be taken to attain particular goals or prevent certain technological consequences. While governance systems are far from ideal at the moment, they are a start in the right direction. AI technology might be used to improve the governance of sustainable development.

The conceptual framework used in this article distinguishes between direct and indirect affects, and as a result, it considers whether an influence on one goal has repercussions on other goals. The term "direct effects" refers to how using AI may have a direct influence on the SDG, whereas "indirect impact" refers to how AI may

[18] Ravallion (2016), pp. 139–64.

[19] Natural Capital Partners (2021). "Industry, Innovation and Infrastructure." https://www.naturalcapitalpartners.com/sustainable-development-goals/goal/industry-innovation-and-infrastructure.

[20] Foer (2017).

have an impact on one objective, which in turn has repercussions for another. If a goal has ripple effects, it means that actions toward that goal will have an impact on other goals as well. For example, artificial intelligence (AI) may have a direct impact on economic growth (SDG 8), yet this growth may be of a type that exacerbates inequality (SDG 10).

SDG 1 (No Poverty- End Poverty in All Its Forms Everywhere) AI will provide real-time resource allocation through satellite mapping and data analysis of poverty. The strength and opportunity such as Emergence of new technologies in primary and industrial sector across developing countries, Predictive power of machine learning upon satellite and aerial images, Deep Learning with mobile device data as a strong domestic income predictor, Combining digital transaction and property data in regression techniques, AI and digital technologies support government decision-making against economic breach. Identifying how AI advances spread globally to ensure equal development (Le Blanc 2015).[21] Digital labor and outsourcing for employment, Block-chain for transparent and corruption-free government processes and Passive data collection bridged with AI and data analysis for reliable poverty estimations.

Perhaps the most ambitious objective of all is to eradicate all types of poverty, everywhere. It is widely assumed that AI has a positive impact on all SDG1 sub-goals and a negative one on just few of them. While the impacts on SDG1 appear to be remarkable, they are virtually exclusively the indirect effects of achieving SDG8a, which is related to economic growth. As previously said, it is extremely improbable that the economic development fostered by contemporary AI would be sustained, inclusive, and sustainable, and even if it is, it is also highly unlikely that AI will aid in the abolition of poverty. This is especially true when we consider that poverty is both an absolute and relative notion, implying that economic development paired with growing inequality will result in more people living in poverty (Nilsson et al. 2016).[22]

SDG 2 (Zero Hunger- End Hunger, Achieve Food Security and Improved Nutrition and Promote Sustainable Agriculture)
Agriculture efficiency will be expanded through predicative examination from imaging with robotized drones and from satellites, constructing a food framework that can meet the dietary requirements of nine billion individuals in the coming future by 2050, while remaining inside the conveying limit of the planet, will require an enormous effort. Such a food framework is expected to convey various SDGs (Berenberg 2018).[23] Given these interconnected difficulties, AI vows to assist with making food creation and utilization more proficient and sustainable. We offer a

---

[21] Le Blanc (2015), pp. 176–187.

[22] Nilsson et al. (2016), pp. 320–322.

[23] Berenberg (2018). "Understanding the SDGs in Sustainable Investing". Germany: Hamburg. https://www.berenberg.de/files/ESG%20News/SDG_understanding_SDGs_in_sustainable_investing.pdf.

couple of models underneath. AI is progressively being sent with other advanced innovation to empower more proficient food production like more food with less information sources. Artificial intelligence assists farmers with better foresee climate designs, adjust planting dependent on supply and demand market, battle bugs and pests, and streamline inputs such as water, manure, pesticides, and so forth, artificial intelligence can be utilized to analyze chemical and biological systems, and increased proficiency in plant reproducing, biotechnology, and agrochemical revelation (Pelto 1987).[24] With the developing measure of information being produced on farms, AI will be fundamental for farmers to use this information to make better decisions.

Agriculture is not immune to the march of automation, and artificial intelligence (AI) is a vital tool in this evolution. AI applications include self-driving agricultural equipment and technology, farm monitoring and management, and the development of novel genetically modified crops. On the other hand, AI-based agriculture is expensive, and there is a genuine risk that all of the aforementioned applications will make agriculture more effective in the affluent world (meso), while widening the gap between farmers in developing and developed countries (macro). Furthermore, AI has no inherent bias toward more nutritious crops and may be utilized just as readily to design and cultivate lucrative crops, crops that are popular and pleasant but lack nutritional value. Finally, there is no reason to believe that AI-based agriculture would be more sustainable or that increased food production in developed nations will result in the elimination of hunger. For a long time, developed nations have produced more food than they require, but this does not imply that the surplus is divided fairly. Rather, it is burned, and there is no reason to believe that AI will change this.

SDG 3 (Good Health and Well-Being: Ensure Healthy Lives and Promote Well-Being for All at All Ages)
AI improves preventative healthcare programmes and diagnostics, resulting in new scientific advances. Eight billion mobile devices with smart phone cameras, for example, are being used to diagnose heart, eye, and blood problems. Few things rival health and wellbeing as human-oriented sustainability goals (D'Alfonso 2017).[25] Positive effects on health might be indirect and arise, for example, through AI driven innovation and study. Furthermore, some may claim that different self-technologies and self-tracking can contribute to individuals living better lives. AI is also being utilized in therapy, and it has the potential to enhance the mental health of individuals who do not have access to human therapists. Self-tracking and self-quantification, on the other hand, are not always beneficial. Furthermore, social media and the other new venues in which AI plays a significant role appear to worsen rather than enhance mental health, while the data is still limited. Another source of concern is that AI powered fitness equipment and dietary analysis and

[24] Pelto (1987).
[25] D'Alfonso et al. (2017), p. 796.

plans, for example, may readily enhance the health of some while widening the gap between those who have access to such technology and those who do not (Thaler and Sunstein 2008).[26] This SDG demonstrates the complexities of AI's effect, as macro and meso[27] level advantages must be balanced against possible micro and meso level downsides.

SDG 4 (Quality Education- Ensure Inclusive and Equitable Quality Education and Promote Lifelong Learning Opportunities for All)

AI-powered virtualized, intelligent mentors and responsive tailored learning are changing education, increasing participation and boosting outcomes. Big data analysis is increasing graduation rates for low-income and first-generation college students by 30%, detecting warning signs before dropout and allowing for focused interventions. Big Tech is highly involved in education, with extensive research and investment targeted towards conquering the education industry. Learning analytics is one area where AI may aid educational analyses, but AI may also be utilized as a teacher. Current intelligent tutoring systems (ITS) have clear limits, but they have proved their potential, and as technology advances, so will in the area of education, effectiveness is. AI may be able to help us achieve SDG 4 in particular through remote instruction. Education becomes much more inexpensive and possibly highly successful, even in the most distant regions of the planet (Shapiro 2017).[28] This, however, necessitates the development of low-cost ITSs for a wide range of languages and disciplines. If the firms building these systems focus on their home markets, which are extremely profitable, the advantages will expand rather than decrease educational disparities. The potential for AI is undeniably enormous, but it need a break from Big Tech and proprietary and Western focused systems. The necessity to ensure that high-quality educational systems are available to all is demonstrated not just by SDG4, but also by the significant effects that would result from achieving this objective. Better education in the poor countries would have enormous advantages for these nations.

SDG 5 (Gender Equality- Achieve Gender Equality and Empower All Women and Girls)

AI is enabling women for growth and new possibilities by detecting and eliminating gender bias and further automating/augmenting activities. In this perspective, equality is solely connected to gender. However, while contemplating the influence of AI, it is equally important to examine problems of race, ethnicity, sexuality, and so on.[29] To begin with the positive, AI may be claimed to help us to achieve this goal through different breakthroughs that emancipate women from work at home, and so

---

[26] Thaler and Sunstein (2008).

[27] In general, a meso-level analysis indicates a population size that falls between the micro and macro levels, such as a community or an organization. However, meso level may also refer to analyses that are specifically designed to reveal connections between micro and macro levels.

[28] Shapiro (2017), pp. 458–460.

[29] Buolamwini and Gebru (2018), pp. 77–91.

on. Furthermore, it might be claimed that AI and automated decision making would help us to overcome human bias in decisions about labor, funding, politics, life in general and so on. In theory, a computer is devoid of bias; however this does not represent the theory laden nature of data and the inescapable human participation in the creation and use of AI. In terms of the potential negative consequences of AI, numerous writers have written critically on how AI affects the weak, notably women and minorities (Chui et al. 2018).[30] While sub-goal 5.b (using enabling technologies to empower women) appears to imply that AI has a positive influence on SDG5:

a. AI is not a required component of enabling technologies, and
b. The great majority of developers are male (and white), therefore it is naive to expect that AI in context will make women and other minorities more equal and less prone to different types of marginalization and prejudice.

SDG 6 (Clean Water and Sanitation- Ensure Availability and Sustainable Management of Water and Sanitation for All)

The Internet of Things (IoT) sensors feeding AI are anticipating sanitation and consumption trends in order to enhance safe water and sanitation delivery. Water and sanitation systems can undoubtedly be handled more efficiently by AI systems and perhaps enhanced via innovation. These are, however, limited consequences that are mostly contingent on attaining SDG9a (innovation) and SDG9b (Nwana 1990).[31] Furthermore, as water systems get more complex, they become more vulnerable to mistake and assault, and access to AI-powered water systems may lead to an increase in inequality between those who have access and those who do not. In the recent development, the development of Robotic Scavenging has been engaged by different country in order to protect the miserable condition and health of manual scavengers. This innovation has not only helped him but has secured his basic human rights and dignity in this complex social system.

SDG 7 (Affordable and Clean Energy- Ensure Access to Affordable, Reliable, Sustainable and Modern Energy for all)

Artificial intelligence real-time analysis is helping to enhance input and efficiency in green energy. The effects of AI on energy systems are significant, although they are mostly indirect effects related to innovation and infrastructure (Muller 2016).[32] The main benefit appears to be more efficient energy production, planning, and distribution systems, and if properly disseminated to developing countries, these effects may be more significant, as better access to energy would cause ripple effects related to SDG8 and SDG9, improving infrastructure, promoting industry, and thus leading to economic growth (Zuboff 2019).[33]

[30] Chui et al. (2018).

[31] Nwana (1990), pp. 251–277.

[32] Müller (2016).

[33] Zuboff (2019).

SDG8 (Decent Work and Economic Growth- Promote Sustained, Inclusive and Sustainable Economic Growth, Full and Productive Employment and Decent Work for All)

When we take into consideration about automation replacing employment, AI has the potential to boost productivity and be a key driver of economic development. SDG8 will be referred to as 8a (relating to economic growth) and 8b (related to social financial development) (referring to decent work). AI has already proved to be a significant accelerator of economic growth and a creator of value, therefore SDG8a is already classed as high impact (Di Vaio et al. 2020).[34] The knock-on consequences of economic progress are equally critical, as prosperity has the potential to alleviate poverty and starving. However, the development that has accompanied the advent of the Big Tech leaders has not been favorable to achieving such aims. There has been growth, but it has been accompanied with rising inequality, as the top percentage of affluent nations around the globe now hold an unprecedented percentage of total wealth (Pekmezovic 2018).[35] Despite numerous problems with properly measuring inequality, data demonstrate that inequality is a genuine and serious issue, and that both intra-country and inter-country inequality are substantial and persistent (SDG Academy 2020).[36] However, within country inequality is becoming more significant, and class, rather than nationality, is now viewed as a driver of global inequality.

Growth that is conducive to achieving SDG8a must be sustained, inclusive, and sustainable. This suggests that while growth is important in and of itself, the type of growth matters a great deal. Taking this into consideration, it is clear that AI is a factor that both allows and hinders the achievement of SDG8a. In connection to these objectives, the indirect impacts of economic growth will be examined, and the major questions for SDG8a are:

(a) If AI supports growth; and
(b) Whether the growth is sustained, inclusive, and sustainable.

The former has been shown to be correct, whereas the latter is far more questionable. There is minimal evidence that AI supports inclusive growth and the objective of developing nations achieving at least 7% GDP growth each year. The influence of AI on decent employment, SDG8b, has the potential to be significant. If inclusive and sustainable economic development and innovation are accomplished, new employment opportunities are likely to be generated, opening up new avenues for good work (Khakurel et al. 2018).[37] However, it is apparent that AI may undermine the decency of labor and produce micro-level harms, both through workplace surveillance and manipulation and through automation. Employers may use surveillance

[34] Di Vaio et al. (2020), pp. 283–314.

[35] Pekmezovic (2018).

[36] SDG Academy. 2020. "Affordable and Clean Energy", https://sdgacademy.org/goal/affordable-and-clean-energy/ (last updated July 14, 2021).

[37] Khakurel et al. (2018), p. 100.

and manipulation tactics to enhance efficiency and control over employees. Furthermore, peer surveillance has the potential to mimic and magnify such tactics. Work is also becoming more automated as a result of AI. While this method may relieve us of tiresome labor and job in general, it is not the same as providing adequate employment possibilities for everybody. Furthermore, rather than remedies such as citizen salaries or massive welfare governments, automation is likely to upset job prospects and the overall condition for the most vulnerable.

SDG 9 (Industry, Innovation and Infrastructure- Build Resilient Infrastructure, Promote Inclusive and Sustainable Industrialization and Foster Innovation)
New hybrid manufacturing, which incorporates AI, IoT sensors, and 4D printing, is changing sectors and resulting in exponential innovation. SDG9 is made up of several themes, including innovation, infrastructure, and industry. 9a will now denote innovation, 9b will denote infrastructure, and 9c will denote industry. The three parts are unmistakably connected, yet they are also unmistakably separate. Sub-goals 9.5, 9.b, and 9.c are primarily concerned with the need for scientific research and technological skills (9.5), domestic technology development (9.b), and improving access to information and communications technology (ICT) (9.c) (Yigitcanlar and Cugurullo 2020).[38]

While AI has the potential to be essential for both public and private sector innovation, a closer examination of the objective indicates that domestic development and improved access to ICT are far more important. AI as it now exists is created in a small number of nations, and while it may be used globally, the advantages and profits from such applications typically return to the home countries of the big businesses in control of data. Meso-level gains might be coupled with macro-level harms in this case. One significant reason for this is that recent advances in AI have relied heavily on access to ever-increasing quantities of data. Data access is neither fair nor equitable, nor is access to the processing power and infrastructure necessary to benefit from cutting-edge AI. Nonetheless, even if wealth is unevenly distributed, contemporary AI may contribute to broad creativity as well as scientific and technological development, which might possibly benefit everybody in the long term.[39] Taking a cue from neoliberal economics, one might claim that AI has a trickledown effect, even though access to the technology and revenues from AI are now dominated by a small number of governments and corporations.

Scientific advancement is said to have a large influence, with significant ramifications. In fact, the majority of the impact of AI can be attributed to the effects of achieving goal 9a; new technology and insight will enable us to achieve most other goals more effectively, and it thus makes more sense to emphasize AI's contribution to this goal, rather than arguing that AI will in effect contribute positively to just

[38] Yigitcanlar and Cugurullo (2020), p. 8548.
[39] Serrano (2018), pp. 134–154.

about all the goals (Toniolo et al. 2020).[40] While innovation and scientific advancements are beneficial, it is critical to remember that innovation and science in the hands of "bad" is just as much a force for evil as it is a force for good in the hands of the good. Scientific "development" is never neutral, and it must always be judged in light of:

(a) Our societies' aims; and
(b) Specific applications.

As a result, AI has a great potential for both negative and positive influence on SDG9a, and the ripple effects of innovation and scientific development are significant. SDG9 is a cornerstone objective for evaluating AI, and it is critical to discern between the direct implications of AI on all other targets and the indirect consequences of SDG9 achievement. Furthermore, AI innovation in Chinese or American private firms is not, by definition, conducive to the type of innovation alluded to in SDG9, and fast innovation in such organizations can easily worsen and expand the disparity between developed and poor nations. For evaluating these consequences, a macro viewpoint is required.

SDG 10 (Reduced Inequalities- Reduce Inequality Within and Among Countries)

Human augmentation via the use of AI-inspired technologies improves physical capacities and corrects impairments, therefore contributing to a more equitable and inclusive society. The beneficial benefits of AI on SDG10 are most likely to be realized through inclusive economic growth. However, because AI does not appear to encourage such growth in the context, the indirect consequences on SDG10 are most likely negative (Chancel 2019).[41] AI has the potential to worsen disparities between affluent and poor in many sectors, and the risk of AI impeding SDG10 is significant (Floridi et al. 2018).[42] This applies both inside and across nations, as both local disparities and global inequalities are not diminishing, at least not as a result of AI.

SDG 11 (Sustainable Cities and Communities- Make Cities and Human Settlements Inclusive, Safe, Resilient, and Sustainable)

AI of Everything, the digital AI mesh powered by IoT, smart gadgets, and wearable's, is already influencing smart cities and assisting in the creation of sustainable communities. For example, technology-intensive smart cities are the outcome of a UN programme to promote the development of sustainable cities and knowledge sharing among the numerous smart cities across the world. Smart cities are technologically advanced, and digital technologies are used to deliver a variety of urban

[40] Toniolo et al. (2020).

[41] Chancel, Lucas. (2019). "Ten Facts About Inequality in Advanced Economies" WID.world Working Paper 2019/15 https://wid.world/document/ten-facts-about-inequality-in-advanced-economies-wid-world-working-paper-2019-15/.

[42] Floridi et al. (2018), pp. 689–707.

infrastructures (Dignum 2019).[43] Despite the importance of AI, it is crucial to emphasize that many smart city ideas are not inherently AI based, even if they are digital. SDG11, on the other hand, involves more than just highly sophisticated science fiction cities; it is about establishing inclusive, safe, resilient, and sustainable communities. AI can undoubtedly play a role in providing safety through monitoring and predictive policing, among other things. Parts of the AI ethics community, on the other hand, are profoundly worried about the "prison-to-tech pipeline," with recurrent calls for the abolition of technology like as face recognition, which is believed to perpetuate racism and other harmful prejudices now hidden behind the glossy veil upon high tech AI. Furthermore, there is no evidence to support the claim that AI leads to inclusive settlements or that it directly promotes resilience for everyone or sustainability in general. Surveillance and AI, on the other hand, are avidly utilized by the more frightening civilizations of the modern world, where it is used to exert what amounts to total control over individuals, a trend that is scarcely consistent with SDG11 (Truby 2020).[44] While governments may weaponize AI to control their populace, it is also linked to increasing polarization, which is another possible impediment to safe and inclusive settlements.

SDG 12 (Responsible Consumption and Production- Ensure Sustainable Consumption and Production Patterns)
With vertical green farms, artificial intelligence is delivering ideal consumption and production levels, minimizing waste and boosting yields and resource efficiency. Consumption is linked to production, and the connection between the two is the major emphasis of SDG12, allowing consumers to consume sustainably (European Commission 2019).[45] Sustainable industry and general innovation may have an indirect impact on this goal, but the main direct positive impact of AI on this goal would be by enabling people to monitor and track their consumption and the associated environmental consequences, resulting in hopefully sustainable consumption patterns (Anderson 2017).[46]

However, AI enabled devices are proliferating, and the internet of things (IoT) phenomena is inextricably linked to AI and data driven innovation. It is a new approach to anticipate obsolescence for previously long-lasting items, in addition to producing new and previously unforeseen requirements with these new products. While earlier TVs became obsolete when new physical technology rendered them less appealing, "smart" TVs, etc., frequently inform users that new applications, etc., will not work on their TVs and that they must be replaced in order to keep current. The beneficial benefits of AI on sustainable consumption appear to be insignificant,

---

[43] Dignum (2019).

[44] Truby (2020), pp. 946–959.

[45] European Commission (2019). "Responsible Consumption and Production | International Partnerships" July 15, 2021, https://ec.europa.eu/international-partnerships/sdg/responsible-consumption-and-production_en.

[46] Anderson (2017).

whereas the negative consequences as AI is viewed in its socio-technical context appear to be significant.

SDG 13 (Climate Action- Take Urgent Action to Combat Climate Change and Its Impacts)

Climate-related problems and disasters may be predicted using data analysis and climate modelling infused with AI. Climate change is increasingly seen as a genuine and serious danger to human (and animal) livelihood. AI excels at optimizing judgments in extremely complicated situations, therefore it may be claimed that AI might potentially enhance climate policy decision making. Some have even claimed that a limited AI technocracy might be built on the necessity to resist developments that human politicians are unwilling or unable to address properly (Nerini et al. 2019).[47] Climate change may be reduced indirectly through AI-powered innovation, infrastructure, and industry (SDG9). Some have even claimed that a limited AI technocracy might be built on the necessity to resist developments that human politicians are unwilling or unable to address properly. Climate change may be reduced indirectly through AI-powered innovation, infrastructure, and industry (SDG9). However, Big Tech has not appeared to be especially green in terms of encouraging a green transition. AI is utilized to improve the efficiency of oil and gas extraction just as much as it is in the development of windmill parks. These are all indirect consequences, and a more immediate worry is how contemporary AI actually generates growing quantities of emissions as a result of data-intensive machine learning training of algorithm (The Climate Reality Project 2017.).[48] This refers to efforts to build green information technologies, since people are increasingly realizing that widespread usage and manufacturing of technology have major environmental costs (Næss 1989).[49] AI and Big Tech are components of an economic system in which traditional development is the main force, and it is likely that AI will have to be separated from this system before its overall implications can be regarded a substantial beneficial factor in addressing climate change.

SDG 14 (Life Below Water- Conserve and Sustainably Use the Oceans, Seas and Marine Resources for Sustainable Development)

Pattern recognition may be used to track marine-life movement, population numbers, and fishing operations in order to improve marine ecosystem sustainability and prevent illicit fishing. Ocean sustainability involves issues for marine ecosystems and marine resources, as well as biological variety in the seas (Sen 2017).[50] This is an area where AI has long been touted as being significant in terms of boat

[47] Nerini et al. (2019), pp. 1–7.

[48] The Climate Reality Project (2017). "The Alliance for Climate Protection: Five Sustainable Cities Making a Difference for the Planet." https://www.climaterealityproject.org/blog/five-sustainable-cities-making-difference-planet.

[49] Næss (1989).

[50] Sen, Conor. (2017). "The 'Big Five' Could Destroy the Tech Ecosystem". *Bloomberg*. November 15, 2017. https://www.bloomberg.com/opinion/articles/2017-11-15/the-big-five-could-destroy-the-tech-ecosystem.

surveillance and monitoring of fisheries and fish stocks. This might, in principle, avoid another occurrence of the type of overfishing that nearly wiped out the north Atlantic cod populations. Another concern linked to SDG14 is ocean acidification, which is partially caused by a failure to meet SDG13, because acidification is a result of carbon dioxide absorption from the atmosphere and therefore an indirect impact of the causes of climate change. While AI may be useful in monitoring marine ecosystems, the additional advantages of AI in this regard are thought to be minor. Furthermore, because overfishing is one of SDG14's key concerns, the harm presented by AI and monitoring systems in the wrong hands may be greater than the benefits of utilizing AI for good. AI may be used to more efficiently identify and track fish populations, as well as to find and monitor and therefore evade those entrusted with stopping illicit operations at sea.

SDG 15 (Life on Land- Protect, Restore and Promote Sustainable Use of Terrestrial Ecosystems, Sustainably Manage Forests, Combat Desertification, and Halt and Reverse Land Degradation and Halt Biodiversity Loss)
Pattern recognition, game theory, and a wide range of computer science applications may be used to track land-animal movement, population numbers, and hunting activities in order to improve sustainable land ecosystems and prevent unlawful poaching. SDG 15 life on land includes terrestrial ecosystems, land use and degradation concerns, and biodiversity. The main reason for anticipating good effects is improved surveillance and monitoring systems, which will allow us to more effectively identify threatened species and places, as well as mitigate bad trends (Engström and Strimling 2020).[51] In this context, a variety of data sources are significant, including registry data on weather, geology, and wildlife, as well as satellite images, and so on.

However, the influence of AI systems on these aims is presently mostly unknown and speculative, and it should be highlighted that surveillance and monitoring systems may possibly be used to prevent these goals from being achieved. Businesses may discover more efficient methods of exploiting land and natural resources, and poachers, for example, may readily utilise satellite images and Prediction algorithms are being used to better efficiently hunt uncommon animals, therefore exacerbating biodiversity loss.

SDG 16 (Peace, Justice and Strong Institutions- Promote Peaceful and Inclusive Societies for Sustainable Development, Provide Access to Justice for All and Build Effective, Accountable and Inclusive Institutions at All Levels)
SDG16, which focuses on society, access to justice, and effective, responsible, and inclusive institutions, is the most politically focused objective. While it is believed that careful implementation of AI can minimize prejudice and corruption while also facilitating widespread access to e-government, personalized, and responsive intelligent services. AI can keep pace with global cyber threats in ways that were

[51] Engström and Strimling (2020), p. 101396.

previously unthinkable (Sætra 2020).[52] While AI can and does favorably impact the majority of sub-goals, there are numerous reasons to be skeptical of this study. With respect to question as to how can AI lead to just, inclusive, and sustainable institutions and societies, then One may claim that AI can be utilized to ensure inclusiveness and nondiscrimination, however such a position reveals a lack of awareness of the fundamentally problematic issues associated with automated decision making, prejudice, and the difficulties associated with detecting such biases (Solove 2000).[53] One potential method to promote democracy through AI would be to investigate how an AI technocracy may ironically lead to a revived democracy. This, however, would need major changes in our political institutions, and there are few grounds to assume that such solutions would

(a) Work; and
(b) Would not have significant negative repercussions.

Concerning the negative consequences, many claim that AI encourages different types of polarization via processes such as filter bubbles and echo chambers (Bender, 2021).[54] Fake news is also linked to AI based social media platforms. Furthermore, AI allows for more effective manipulation or nudging of humans, and it is doubtful that actors with different motives will refrain from utilizing such tactics to promote their own self-interest above the interests of other citizens, consumers, and persons in general.

SDG 17 (Partnerships for Goals- Strengthen the Means of Implementation and Revitalize the Global Partnership for Sustainable Development)
The United Nations should play a significant role in bringing together governments, business, academia, and civil society to investigate the responsible development of human-centric AI in addressing humanity's issues. The implementation of the SDGs and the promotion of partnerships are areas where AI may play a helpful role, such as monitoring systems for compliance, etc., however this is not commonly believed that AI will play a key role in achieving SDG17 (Danaher 2019).[55] AI, on the other hand, can play a critical role as the subject of both regulations and policy for the Partnership for Sustainable Development and, in particular, as a technology that must be transferred and made more widely available for all in order to achieve just about any of the SDGs that involve capacity building in developing countries and the promotion of equality and fairness (which is, in fact, the case). The system in which data intensive AI is almost entirely controlled by a few number of strong nations is a major impediment to the SDGs in general, and hence a major source of worry for those working toward SDG17.

[52] Sætra (2020), p. 101283.
[53] Solove (2000), p. 1393.
[54] Bender et al. (2021).
[55] Danaher (2019).

The 17 Sustainable Development Goals covers all global issues which every country is currently facing at international as well as national level. The above analysis shows that AI can be utilized as an effective tool in order to achieve the expected goals. It has been examined that AI is directly affecting several goals whereas other goals are indirectly affected by it. AI would have negative impact on several goals but the positive aspect of this new-age technology cannot be ignored. Since it is an era of industrial revolution 4.0, every state and its machinery should acknowledge such technological development and should utilize and adopt it within its domestic framework. There is a dare need to frame an effective policy regarding the employment of AI and its entities in sustainable development goals and the realization of social, economic, and environment progress can be done through the growth of AI. These ideal goals would be mere illusion if the global community would not focus on the implementation and application of such innovative technologies in various sectors.

5 Conclusion

Finally, Artificial intelligence (AI) is quickly expanding into new sectors such as commerce, corporate operations, and government policy. Machine intelligence and robots with deep learning capabilities have had significant disruptive and enabling effects on businesses, governments, and society. Therefore, preliminary findings from sustainable AI development indicate that there is a great deal of opportunity to think about AI in terms of sustainable development. AI, being a collection of general-purpose technologies, does not have a single future. It has multiple futures, and these futures are dependent on factors other than technical advancements. They are dependent on decisions made to support research and innovation, decisions made to deploy technology in society, and decisions made on how to design and execute technologies. AI applications have the potential to play a significant role in society, but we must determine which role they will play. Using sustainable AI development as a framework makes us aware of our options. They're also having an impact be it positive impact or negative impact upon each SDGs global trend. As the AI revolution alters our society, it may signal a utopian future in which humans and machines coexist peacefully, or it may herald a dismal future filled with conflicts, poverty, and pain. More immediately, study shows whether AI help us achieve the Sustainable Development Goals (SDGs) of the United Nations (UN) or push us farther down the road to more economic and environmental uncertainty, ecological breakdown and social upheaval. This article provides a broad perspective and thought on the role and influence of AI and its related technologies in achieving the Agenda 2030 for Sustainable Development, namely its 17 Sustainable Development Goals (SDGs). An analysis into all the challenges and opportunities for progress led by these technologies, under different dimensions of perspectives of human needs underlying the SDG, based on an in-depth analysis of relevant literature portrays the nexus between AI technologies and each of the SDGs. A final discussion of essential recommendations or priorities to follow in order to ensure good development toward the aforementioned goals by 2030 is looked after with all the positive and negative impact the AI will be having upon all the SDGs.

References

Anderson E (2017) Private government: how employers rule our lives (and why we don't talk about it). Princeton University Press, Princeton

Bender EM, Gebru T, McMillan-Major A, Shmitchell S (2021) On the dangers of stochastic parrots: Can language models be too big. In: Proceedings of the 2020 conference on fairness, accountability, and transparency. Association for Computing Machinery, New York

Berenberg (2018) Understanding the SDGs in Sustainable Investing. Hamburg, Germany. https://www.berenberg.de/files/ESG%20News/SDG_understanding_SDGs_in_sustainable_investing.pdf

Borukhovich E (2015) How Will Artificial Intelligence Change Healthcare? https://www.weforum.org/agenda/2015/10/how-will-artificial-intelligence-change-healthcare (last updated July 15, 2021)

Buolamwini J, Gebru T (2018) Gender shades: Intersectional accuracy disparities in commercial gender classification. In: Conference on Fairness, Accountability and Transparency, 77–91. http://proceedings.mlr.press/v81/buolamwini18a.html

Chancel L (2019) Ten Facts About Inequality in Advanced Economies. WID.world Working Paper 2019/15 https://wid.world/document/ten-facts-about-inequality-in-advanced-economies-wid-world-working-paper-2019-15/

Chui M, Harryson M, Manyika J, Roberts R, Chung R, van Heteren A, Nel P (2018) *Notes from the AI Frontier: Applying AI for Social Good*. McKinsey Global Institute, New York

D'Alfonso S, Santesteban-Echarri O, Rice S, Wadley G, Lederman R, Miles C, Gleeson J, Alvarez-Jimenez M (2017) Artificial intelligence-assisted online social therapy for youth mental health. Front Psychol. 8:796

Danaher J (2019) Automation and Utopia: Human Flourishing in a World without Work. Harvard University Press, Cambridge

Dignum V (2019) Responsible artificial intelligence: how to develop and use ai in a responsible way. Springer, Cham

Engström E, Strimling P (2020) Deep learning diffusion by infusion into preexisting technologies—Implications for users and society at large. Technol Sociol 63:101396

European Commission (2019.) Responsible Consumption and Production | International Partnerships July 15, 2021, https://ec.europa.eu/international-partnerships/sdg/responsible-consumption-and-production_en

Floridi L, Cowls J, Beltrametti M, Chatila R, Chazerand P, Dignum V, Luetge C, Madelin R, Pagallo U, Rossi F, Schafer B, Valcke P, Vayena E (2018) AI4People—An ethical framework for a good AI society: Opportunities, risks, principles, and recommendations. Minds Mach 28: 689–707

Foer F (2017) World without mind. Random House, New York

International Institute for Sustainable Development (2020) Sustainable Development | International Institute for Sustainable Development. Last updated July 12, 2021. https://www.iisd.org/about-iisd/sustainable-development

Jadhav N (2019) New-Age Technology and Industrial Revolution 4.0 : Global Public Policy Issues in Economy, Democracy, National Security and World Peace. Konark Publishers Pvt Limited

Khakurel J, Penzenstadler B, Porras J, Knutas A, Zhang W (2018) The rise of artificial intelligence under the lens of sustainability. *Technologies* 6(4):100

Le Blanc D (2015) Towards integration at last? The sustainable development goals as a network of targets. Sustain Dev 23:176–187
Lideniro A (1991) Potential applications for artificial intelligence in the petroleum industry. J Petroleum Technol 43:1306–1309. https://doi.org/10.2118/21138-PA
Müller VC (ed) (2016) Risks of artificial intelligence. CRC Press, Boca Raton
Næss ADE (1989) Ecology, community and lifestyle: outline of an ecosophy. Cambridge University Press, Cambridge
Natural Capital Partners (2021) Industry, Innovation and Infrastructure. https://www.naturalcapitalpartners.com/sustainable-development-goals/goal/industry-innovation-and-infrastructure
Negnevitsky M (2005) Artificial intelligence a guide to intelligent systems artificial intelligence artificial intelligence a guide to intelligent systems. Pearson Ltd, United Kingdom
Nerini FF, Sovacool B, Hughes N, Cozzi L, Cosgrave E, Howells M, Tavoni M, Tomei J, Zerriffi H, Milligan B (2019) Connecting climate action with other sustainable development goals. Nat Sustain:1–7
Nilsson M, Griggs D, Visbeck M (2016) Policy: Map the interactions between sustainable development goals. Nature 534:320–322
Nwana HS (1990) Intelligent tutoring systems: an overview. Artif Intell Rev 4:251–277
Palomares I, Martínez-Cámara E, Montes R, García-Moral P et al (2021) A panoramic view and swot analysis of artificial intelligence for achieving the sustainable development goals by 2030: progress and prospects. Appl Intell 51:6497–6527. https://doi.org/10.1007/s10489-021-02264-y
Pekmezovic A (2018) The UN and goal setting: From the MDGs to the SDGs. In: Walker J, Pekmezovic A, Walker G (eds) Sustainable development goals: harnessing business to achieve the SDGs through finance, technology, and law reform. John Wiley & Sons Ltd., UK
Pelto PJ (1987) The snowmobile revolution: technology and social change in the arctic. Waveland Press Inc, Long Grove
Ravallion M (2016) Are the world's poorest being left behind? J Econ Growth 21:139–164
Sachs JD (2012) From millennium development goals to sustainable development goals. Lancet 379:2206–2211
Sætra HS (2020) A shallow defence of a technocracy of artificial intelligence: examining the political harms of algorithmic governance in the domain of government. Technol Soc 62: 101283
Sætra HS (2021) Ai in context and the sustainable development goals: factoring in the unsustainability of the sociotechnical system. Sustainability 13:1–19
SDG Academy (2020) Affordable and Clean Energy. https://sdgacademy.org/goal/affordable-and-clean-energy/ (last updated July 14, 2021)
Sen C (2017) The 'Big Five' Could Destroy the Tech Ecosystem. Bloomberg. November 15, 2017. https://www.bloomberg.com/opinion/articles/2017-11-15/the-big-five-could-destroy-the-tech-ecosystem
Serrano W (2018) Digital systems in smart city and infrastructure: digital as a service. Smart Cities 1:134–154
Shapiro A (2017) Reform predictive policing. Nature News 541:458–460
Shield N (2019) Amazon Is Introducing New Warehouse Robots. https://www.businessinsider.com/amazon-introduces-new-warehouse-robots-2019-5?IR=T
Solove DJ (2000) Privacy and power: computer databases and metaphors for information privacy. Stanford Law Rev 53:1393
Spector L (2006) Evolution of Artificial Intelligence. Artif Intell 170:1251–1253. https://doi.org/10.1016/J.ARTINT.2006.10.009
Thaler RH, Sunstein CR (2008) Nudge: improving decisions about health, wealth, and happiness. Yale University Press, New York

The Climate Reality Project (2017) The Alliance for Climate Protection: Five Sustainable Cities Making a Difference for the Planet. https://www.climaterealityproject.org/blog/five-sustainable-cities-making-difference-planet.

The World Bank (2015) Millennium Development Goals. https://www5.worldbank.org/mdgs/

Toniolo K, Masiero E, Massaro M, Bagnoli C (2020) Sustainable business models and artificial intelligence: opportunities and challenges. In: Knowledge, people, and digital transformation. Springer, Germany Berlin

Truby J (2020) Governing artificial intelligence to benefit the UN sustainable development goals. Sustainability Division 28:946–959

UN Department of Public Information (2015) We Can End Poverty, 2015. https://www.ndi.org/sites/default/files/Handout 7 - Millennium Development Goals.pdf

United Nation Development Program (2020) Sustainable Development Goals. https://www.africa.undp.org/content/rba/en/home/sustainable-developmentgoals/background.html#:~:text=The%20Sustainable%20Development%20Goals%20(SDGs,economic%20challenges%20facing%20our%20world

United Nations (2005) Department of Economic and Social Affairs| Sustainable Development. Last updated July 12, 2021 https://sdgs.un.org/goals

Vaio D, Assunta RP, Hassan R, Escobar O (2020) Artificial intelligence and business models in the sustainable development goals perspective: a systematic literature review. J Bus Res 121:283–314

Vinuesa R et al (2020) The role of artificial intelligence in achieving the sustainable development goals. Nat Commun 11:1–10

Yigitcanlar T, Cugurullo F (2020) The sustainability of artificial intelligence: An urbanistic viewpoint from the lens of smart and sustainable cities. Sustainability 12:8548

Zuboff S (2019) The age of surveillance capitalism: the fight for a human future at the new frontier of power: Barack Obama's Books of 2019. Public Affairs, New York

Sustainable Growth and the Role of Artificial Intelligence in Improving the Circular Economy

Carlo Amatucci and Giovanni Mollo

Abstract Sustainable development has three dimensions, economic, social and environmental. The three-fold dimension of sustainable development implies the consideration of the environmental issue that goes beyond mere protection of the environment since it cannot be considered separately from the economic-productive system and social needs. The circular economy operates along the same three dimensions. A central role in the successful implementation of the circular economy must necessarily be given to the enterprises. Artificial intelligence provides a powerful tool in the transition to a circular economy, acting as an instrument for innovation in the business model, by directing business activity towards sustainable capitalism, overcoming the model of capitalism aimed solely at profit.

Keywords Circular economy · Artificial intelligence · Sustainable business model

1 Introduction

Sustainable development has an economic, social, and environmental dimension. This three-fold dimension implies the consideration of the environmental issue that goes beyond mere protection of the environment since it cannot be considered separately from the economic-productive system and social needs. The circular

The article was previously published in the journal Law&Digital Technologies, 2022, 2(1): 7–15. Although the article is a joint work, paragraphs 1 and 2 are attributable to Giovanni Mollo and paragraphs 3 and 4 to Carlo Amatucci. The opinions expressed are personal and in no way involve Consob.

C. Amatucci (✉)
The University of Naples Federico II, Napoli, Italy
e-mail: amatucci@unina.it

G. Mollo
Consob, Rome, Italy

L. V. Sannikova (ed.), *Digital Technologies and Distributed Registries for Sustainable Development*, Law, Governance and Technology Series 64,
https://doi.org/10.1007/978-3-031-51067-0_2

economy can play a crucial role in the fulfilment of sustainable growth, mainly to help tackle the climate crisis by reducing GHG emissions along supply chains.

Artificial intelligence provides a powerful tool in the transition to a circular economy, acting as an instrument for the innovation of business activities towards sustainable capitalism, and overcoming the model of capitalism aimed at profit maximization only. Such a desirable path will require a new enterprise definition based on the concept of the "Sustainable business model".

2 Sustainable Development

The principle of sustainable development has its roots in the United Nations Conference on the Human Environment held in Stockholm in 1972. In fact, the first principle proclaimed as a result of the work of the Conference, states that "man has a fundamental right to freedom, equality and satisfactory living conditions in an environment which allows him to live in dignity and well-being, and he bears a high degree of responsibility for the protection and improvement of the environment for future generations", and the second principle states that "the Earth's natural resources, including air, water, flora, fauna and especially the natural ecological system, must be safeguarded for the benefit of present and future generations through careful planning or appropriate management".[1]

The principle was reformulated in paragraph 27 of the 1987 Report of the World Commission on Environment and Development of the United Nations, entitled Our Common Future, which states that "Humanity has the ability to make development sustainable to ensure that it meets the needs of the present without compromising the ability of future generations to meet their own needs".[2]

This principle was reaffirmed by the United Nations Conference on Environment and Development, held in Rio de Janeiro from 3 to 14 June 1992, according to which "human beings are at the center of concerns for sustainable development. They have the right to a healthy and productive life in harmony with nature".[3] It was also stated that "the right to development must be realized in such a way as to equitably meet the environmental and developmental needs of present and future generations".[4]

At the 2002 UN Summit on Sustainable Development in Johannesburg, the participants solemnly agreed on the following declaration: "We firmly pledge to

[1]United Nation, *The Stockholm Declaration* 1972, https://www.un.org/en/conferences/environment/stockholm1972.

[2]Federal Office for Spatial Development ARE. *Brundtland Report* 1987. https://www.are.admin.ch/are/it/home/sviluppo-sostenibile/cooperazione-internazionale/agenda2030/onu-_-le-pietre-miliari-dello-sviluppo-sostenibile/1987%2D%2Drapporto-brundtland.html.

[3]United Nation, *The Rio Declaration on Development and the Environment* 1992, https://www.un.org/en/development/desa/population/migration/generalassembly/docs/globalcompact/A_CONF.151_26_Vol.I_Declaration.pdf.

[4]Ibid.

use our rich diversity, which is our collective strength, in constructive partnerships for change and for the achievement of the common goal of sustainable development", and "we agree on the need for the private sector to operate within transparent and stable rules to strengthen corporate public accountability".[5] In 2012, the UN organized in Rio de Janeiro the United Nations Conference on Sustainable Development (UNCSD), also called Rio+20, at the end of which the final document introduced the idea of a green economy in a context of sustainable development and poverty reduction.[6]

In 2015, the UN Conference on Climate Change, held in Paris, culminated with the adoption of the 2030 Agenda for Sustainable Development, which consists of 17 objectives (OSS/SDGs, Sustainable Development Goals) and 169 targets.[7] The goals and targets, as stated in the preamble of the 2030 Agenda, "are interconnected and indivisible and balance the three dimensions of sustainable development: the economic, social and environmental dimensions".[8] This statement is the key to the proper implementation of the goals and targets set by Agenda 2030.

The principle of sustainable development became part of the Italian legal system with art. 3-quater of Legislative Decree 152/2006 and subsequent amendments, the so-called Environmental Code, which states that all human activity legally relevant under this code must comply with the principle of sustainable development, in order to ensure that the satisfaction of the needs of current generations cannot compromise the quality of life and opportunities of future generations. The principle of sustainable development becomes central in the action of the public administration pursuant paragraph 2 of the aforementioned provision, according to which in the comparative choice of public and private interests characterized by discretion, the interests of environmental protection and cultural heritage must be given priority consideration.

3 Sustainable Development and the Socio-Economic System

Sustainable development has three dimensions: economic, social and environmental.[9] Its goals and targets are the result of the balancing of these three dimensions; therefore, they are interconnected and indivisible.

---

[5] United Nation, *The Johannesburg Declaration on Sustainable Development. From our origins to the future* 2002, https://www.un.org/esa/sustdev/documents/WSSD_POI_PD/English/POI_PD.htm.

[6] United Nation, *The further we want* 2012, https://www.minambiente.it/sites/default/files/archivio/allegati/rio_20/the_future_we_want_english.pdf.

[7] United Nation, *The Paris Agreement* 2015, https://unfccc.int/process-and-meetings/the-paris-agreement/the-paris-agreement.

[8] Ibid.

[9] Fraccia (2010b, 2012); Freeman et al. (2018); Greco and Salimbeni (2003); Lanza (2002); Lomonaco (2004); Pepe (2002a), pp. 209–243. Pepe (2002b).

The existence of a three-dimensional space in which sustainable development takes place implies a consideration of the environmental issue that goes beyond the mere protection of the existing environment, since environmental protection cannot be considered separately from the economic and productive system and from social needs.

Earlier research pointed out that the protection of the existing environment, that did not consider the needs of development, would have undermined "the ultimate purpose of protection of the human species that inspires the environmental discipline, because, frustrating, beyond a certain limit, the development needs of the current ones, would lose sight of the interests of future generations".[10] Following the same thought, by definition, the problem of environmental protection arises only in the presence of development. Pursuing only environmental protection would result in a situation of conflict with sustainable development "which should be resolved in favor of the second since the (true) environmental protection is justified only because of (or if compatible with) sustainable development".[11]

The principle of sustainable development, however, overcomes the aforementioned opposition as environmental protection represents one of the three dimensions of sustainable development.

The environmental challenge, in fact, "inevitably raises questions that also (and above all) concern the way the economic and productive system functions and is structured".[12] The environmental issue has imposed the search for new development models able to combine "the increasingly challenging objectives of environmental protection with the characteristics of market economies".[13]

In this context, the idea of a Green economy was introduced, with an uncertain progress.[14] The concept of Green economy was established within the United Nations, as a system capable of ensuring economic development, generating long-term benefits without, however, exposing future generations to significant environmental risks.[15]

The UNEP (UN Environment Program), in its report on the global green economy, defined the Green economy as "an economy that results in improved human well-being and social equity, while significantly reducing environmental risks and the destruction of non-renewable resources. It is a low-carbon, resource-efficient and socially inclusive economy".[16] This definition led to the assumption that there was a direct connection between the Green economy and sustainable development. "The green economy would be the means to enable the simultaneous advancement of

[10] Fraccia (2010a), p. 15.

[11] Ibid. 20.

[12] Moliterni (2020), p. 32.

[13] Ibid. 33.

[14] UNDESA (2012).

[15] Moliterni (2020), p. 33.

[16] Fortunato (2019), p. 420.

environmental and economic goals according to the logic of the same concept of sustainable development".[17]

The interpretation, according to which the Green economy is linked to sustainable development by a relationship of instrumentality, extends the problems and doubts of interpretation inherent to sustainable development to Green economy. In particular, the interpretation, according to which sustainable development represents a formula for a simple composition of the interests involved, gives prevalence to the economic aspect and to the consideration of the environmental aspect "as one of the elements to be included in the economic choices (and not as the prerequisite to rethink the overall structure of the system)".[18] This approach reduced the Green economy to the marginal role of "container within which a plurality of 'measures' and intervention tools can be brought together", united only by being instrumental to the reduction of polluting emissions and exploitation of natural resources.[19] In this way, the economic system would continue to be structurally independent of the environmental variable, the latter being relegated to operate within the limits of its "green part" identified in the Green economy.[20] The Green economy would therefore be at least insensitive to the social dimension of development and, in particular, to the question of access to public goods and collective resources.[21] Hence, the hypothesis of a passage from the Green economy to the Ecological economy,[22] characterized by a close connection between the functioning of the ecological system and the economic system, and, more recently, the overcoming of the Green economy in favor of the Blue economy,[23] abandoning the linear economy (red economy)[24] in favor of the circular economy (blue economy).[25] The latter represents "a real paradigm shift that imposes significant changes in the use of natural resources, waste management, design and life cycle of goods".[26]

---

[17] Moliterni (2020), p. 34.

[18] Ibid.

[19] Ibid. 35.

[20] Ibid.

[21] Moliterni (2020), p. 37.; Goodman and Salleh (2013), pp. 411–424; Mattei (2011).

[22] Munda (1997), pp. 213–233.

[23] Pauli (2015).

[24] De Leonardis (2017), pp. 163–207.

[25] Moliterni (2020); Ferrara (2018), pp. 801–835.

[26] Cocconi (2020).

4 Sustainable Development and Circular Economy

Circular economy[27] is an alternative to our current *linear economy* based on the assumption that our planet has endless *resources.*[28] It's an economy designed to be *regenerative* and *restorative*, keeping "*products, components and materials at their highest utility and value, at all times*",[29] where products, components and materials are designed and made for reuse, refurbishment, recycling and returning materials to the planet. The Ellen Mc Arthur Foundation, one of the world most active partner in the promotion of the circular economy, declared that "our way of doing things is reaching its limits. The current system is no longer working for businesses, people or the environment. We take resources from the ground to make products, which we use, and, when we no longer want them, throw them away. Take-make-waste. We call this a linear economy. The linear economy has to change. We must transform all the elements of the take-make-waste system".[30] In the circular economy "natural systems are regenerated, energy is from renewable sources, materials are safe and increasingly from renewable sources, and waste is avoided through the superior design of materials, products, and business models. A circular economy offers a positive way forward by redefining value creation to focus on society-wide benefits. It addresses the shortcomings of the current system, while creating new opportunities for businesses and society. Circular economy principles present unique opportunities to help tackle the climate crisis by reducing GHG emissions along supply chains"[31] (Ellen MacArthur Foundation 2019).

The basic idea under the circular economy model is that economic models should be inspired by the way natural ecosystems work "where everything is reused and nothing is lost".[32] The economic system should mimic the natural one, thus recognizing "the existence of insurmountable limits that impose the continuous reuse of goods",[33] the idea known in the literature as biomimicry.

The Circular economy is, in fact, based on Ecological Economics, which considers economic activity as an extension of biological activity.[34]

[27] Cocconi (2020); De Leonardis (2017); De Leonardis (2019); De Leonardis (2020), pp. 50–72.; Bovino (2014), pp. 682–691.; Scotti (2019), pp. 493–529.

[28] On the origins of the circular economy, F. De Leonardis (2017) attributes its first formulation, in 1970, to architect J. T. Lyle, with "regenerative design" based on the use of "local and renewable resources", and to architect W. Stahel who argued that the linear economic production model (take, produce, use and throw away) was not sustainable, due to increased demand for raw materials and waste accumulation.

[29] Webster (2017).

[30] Ellen MacArthur Foundation. *Completing the picture. How the circular economy tackles climate change.* 2019. https://emf.thirdlight.com/link/dcijanpohgkd-oblthh/@/preview/5.

[31] Ibid.

[32] De Leonardis (2017), p. 168; Cocconi (2020), p. 17.

[33] De Leonardis (2017), p. 168.

[34] De Leonardis (2017), p. 180.

Circular economy represents, therefore, a synthesis between economics and ecology (bioeconomy), which tends to optimize the value of goods and, therefore, operates on their life cycle, taking care not only for "the end of life of the products, but also their beginning of life"[35] (De Leonardis 2017, p. 182; Cocconi 2020, p. 30). Following such model, the production process should not simply resolve the problem of how to use what remains at the end of the life of the product, but it should require the design of what nature offers "to use it and return it to nature".[36]

The issue is no longer the proper management of waste or the use of waste as raw materials.[37] According to the circular economy, the production process starts from agriculture, that is, from the use of locally available resources and it returns to agriculture. Waste is only a transitory phase on the way back to nature.[38]

The bio-economy, which seems to have only two of the three dimensions of sustainable development, i.e. economic and environmental, actually contains the third one, i.e. social. The unsustainability of the system, as it has been pointed out, does not consist in the depletion of non-renewable resources.[39] On the contrary, the critical factor is the geographical location of non-renewable raw materials, which are concentrated only in certain areas of the planet and constitute reserves that are not easily accessible, thus generating "problems in terms of geopolitical security".[40] Circular economy, in its most evolved version, operates along the three cited dimensions of sustainable development and involves all its players, such as public authorities, citizens, businesses, and collective subjects: "The transition to a circular economy requires the participation and commitment of different groups of people. The role of policymakers is to provide structural conditions, predictability and trust for businesses, to enhance the role of consumers, and to define how citizens can benefit from the changes taking place. The business community can completely redesign supply chains, targeting resource efficiency and circularity.".[41] All the actors involved are equally important, but a particularly delicate role is played by Company Law, not only as the engine of business and production but also because of its links with both consumers and financiers.

The circular economy is high on Europe's political agenda. On March 11, 2020, the Commission issued a Communication titled "A new circular economy Action Plan for a cleaner and more competitive Europe". The Plan presents "a set of

[35]De Leonardis (2017), p. 182.; Cocconi (2020), p. 30.

[36]De Leonardis (2017), p. 183.

[37]Ibid. 179.

[38]Ibid. 183.

[39]Cocconi (2020), p. 18.

[40]Ibid.

[41]European Commission (2014). *The Circular Economy. Connecting, Generating and Retaining Value.* https://op.europa.eu/en/publication-detail/-/publication/c8cfd1ae-6285-40ba-879f-f2e78e4c2b6e.

interrelated initiatives to establish a strong and coherent product policy framework that will make sustainable products, services and business models the norm".[42]

5 Artificial Intelligence and Circular Economy

A central role in the implementation of the circular economy must necessarily be attributed to the business enterprise.[43] At the helm of the transition to a circular economy is the business community. Without considering the current regulatory principles and constraints that govern the various business models, enterprises should consider the economic, social and environmental dimensions in which they operate to fulfill two principles well defined by the visionary outlooks of Evan and Freeman (1988) more than thirty years ago. The first one (principle of corporate rights) states that the corporation and its managers should not violate the legitimate rights of others to determine their own future. The second one (principle of corporate effects) postulates that the corporation and its managers are responsible for the effects of their actions on others.

The social dimension requires the corporation to consider not only the demands coming from its shareholders and from various components which even indirectly are involved in the operations but also the demands coming from society in general.

The environmental dimension requires managers to organize their production cycle in accordance with the principles mentioned above, which characterize the circular economy in its most advanced form.

As already pointed out by the European Commission, socially responsible conduct for companies implies—going beyond the full compliance of their legal obligations—an additional investment in human capital, in the environment and in the relationships with other interested parties[44] (European Commission 2014). Even if the main purpose of companies is to make a profit—a premise which, now more than ever, is questioned by prominent scholars worldwide and by some jurisdiction as we will see[45]—those companies can also contribute to social and environmental objectives by integrating social responsibility as a strategic investment into their business strategy, management tools and operations. In order to act in a socially responsible manner, companies should integrate social, environmental, ethical, human rights and consumer concerns into the definition of their business strategies in order to "create shared value among their owners/shareholders and other

[42] European Commission (2020a) *A new Circular Economy Action Plan.* https://eur-lex.europa.eu/legal-content/EN/TXT/?qid=1583933814386&uri=COM:2020:98:FIN.

[43] European Commission (2014).

[44] European Commission (2014).

[45] Sjåfjell and Bruner (2020); Strine (2019); Freeman et al. (2018); Millon (2013), pp. 1013–1044.; Stout (2012); Keay (2011).

stakeholders and society at large; identify, prevent and mitigate their possible adverse effects" (European Commission 2011).[46]

The challenge to which companies are called can be facilitated, even if not in all the objectives that are defined in the 2030 Agenda,[47] by the use of new technologies and, in particular, by the use of Artificial Intelligence.[48] The European Commission itself has, in fact, highlighted that "the use of AI systems can play a significant role in achieving the goals of sustainable development and in supporting the democratic process and social rights".[49] The European Commission considers the application of the data economy, and in particular AI, as an opportunity of growth for Europe itself, believing that it has the tools "to become a world leader in innovation in the data economy and its applications".[50] The use of artificial intelligence, ensuring improved forecasting, the optimization of operations and resource allocation, and the personalization of service delivery, can contribute to the achievement of results socially and environmentally beneficial as well as provide competitive advantages fundamental to European businesses and the economy.[51] However, the same elements and techniques that fuel the socio-economic benefits of AI may also entail new risks or negative consequences for individuals or society.[52] In view of the speed of technological change and possible challenges, the EU is committed to pursuing a balanced approach by formulating a proposal to regulate this phenomenon.[53]

The European Commission's White Paper on Artificial Intelligence has already laid out the policy options on how to achieve the twin goals of promoting the adoption of AI and address the risks associated with certain uses of this technology. The regulatory proposal aims to implement the second objective in order to develop a trusted ecosystem by providing a legal framework for trusted AI. The proposal is based on the fundamental values and rights of the EU and aims to give people and other users the confidence to adopt solutions based on AI, while encouraging businesses to develop them.

---

[46] European Commission (2011). *A renewed EU Strategy 2011-14 on Corporate Social Responsibility*. https://eur-lex.europa.eu/LexUriServ/LexUriServ.do?uri=COM:2011:0681:FIN:EN:PDF.

[47] Vinuesa et al. (2020), p. 233.

[48] Joshi (2019).

[49] European Commission (2020b) *White paper on artificial intelligence - A European approach to excellence and trust.* https://ec.europa.eu/info/publications/white-paper-artificial-intelligence-european-approach-excellence-and-trust_en.

[50] European Commission (2020b).

[51] Proposal for a Regulation of the Parliament and of the Council laying down harmonised rules on artificial intelligence act and amending certain Union legislative acts, 21.04.2021, p. 1, https://eur-lex.europa.eu/resource.html?uri=cellar:e0649735-a372-11eb-9585-01aa75ed71a1.0001.02/DOC_1&format=PDF.

[52] Ib.

[53] Abriani Niccolò (2022), p. 761. In this regard, the author evokes "Corporate Digital Responsibility, understood as a set of practices and behaviors that help complex organizations, such as large enterprises, use data and technology responsibly, including socially and environmentally, as well as economically and technologically".

In particular, the proposed regulation has the following specific objectives: "to ensure that AI systems placed on the Union market and used are secure and comply with existing fundamental rights legislation and Union values; to provide legal certainty to facilitate investment and innovation in artificial intelligence; to improve the governance and effective enforcement of existing legislation on fundamental rights and security requirements applicable to AI systems; to facilitate the development of a single market for lawful, secure, and reliable AI applications reliable as well as prevent market fragmentation".[54]

Its concrete implementation is a development of a pan-European network of supercomputers, capable of competing at continental level, consisting of eight units (one of which has been recently installed in Bologna).[55] In particular, companies would take advantage of new generations of products and services in which Europe is particularly strong as the green and circular economy.[56]

Artificial intelligence is an essential tool in retrieving and processing large amounts of data in the process of transition to a circular economy.[57] In this case, AI acts as an instrument for innovating the business model,[58] directing company activity towards sustainable capitalism, overcoming the capitalism model aimed at gaining just profit.[59]

"Artificial intelligence is a fast evolving family of technologies that can contribute to a wide array of economic and societal benefits across the entire spectrum of industries and social activities".[60] Several definitions have been given of AI, but all of them expressed the idea of creating "computer programs or machines capable of behaviors that we would consider intelligent if enacted by humans. In 1955, John McCarthy, one of the founding fathers of the discipline, described the process as "consisting of getting a machine to behave in ways that would be called intelligent if it were a human being behaving that way".[61] In essence, these are systems, governed by an algorithm of machine learning, which are fed by the so-called big data, that is, the data generated from the set of information collected through digital devices. The systems process and correlate the information thus collected according to the rules

[54] Proposal for a Regulation of the Parliament and of the Council laying down harmonised rules on artificial intelligence act and amending certain Union legislative acts, 21.04.2021, p. 3.

[55] Cineca. n.d. https://www.cineca.it/temi-caldi/Leonardo.

[56] Ibid.

[57] Enriques and Zetzsche (2020), pp. 55–98.; Montagnani (2020), p. 1003.

[58] Di Vaioa et al. (2020), pp. 283–314.

[59] Montagnani (2020); Accenture. (2017). *Boost Your AIQ Transforming into an AI Business.* https://newsroom.accenture.com/news/artificial-intelligence-leaders-generate-greater-shareholder-value-finds-new-report-from-accenture-research.htm.

[60] Proposal for a Regulation of the Parliament and of the Council laying down harmonised rules on artificial intelligence act and amending certain Union legislative acts.

[61] Kaplan (2018).

imposed by the programmer with the algorithm.[62,63] The elaborations and the correlations of the collected information, according to the rules imposed to the system with the algorithm, constitutes the result of a process that is self-feeding and able to create new models starting from initial data.[64] In other words, these systems are structurally capable of adapting to the data they receive, creating new models that allow the system to perform its task more efficiently.[65] This means that the machine is able to learn, to adapt to new conditions that arise and which are different from the initial ones.[66,67] Of particular interest here are the predictive algorithms, "which detect prospective correlations useful to orient strategic choices on the basis of what has been learned from the data".[68]

In this sense, AI can therefore become a fundamental tool for creation value, "thanks to its unquestionable ability to refine the predictive models needs to offer more accurate and in-depth information to directors and top managers".[69]

The AI, therefore, normally constitutes a tool offered to company directors for the fulfillment of their mandate and this fact, as said, requires a reorganization of the business model.[70]

Combining the power of AI with a vision for a circular economy means a significant opportunity to reshape the world economy. From electronics to food production, there are several potential applications of artificial intelligence that will help to speed up the transition to the circular economy. Products at the end of their life are hard to automatically disassemble, sort, and separate. Robotic sensors recognize and sort different waste products for recycling.

In the food industry, AI enhances opportunities in farming, processing, logistics and waste management. It is already being used to scan fruit; from the images produced AI can determine the optimum time to pick them. Italy, for instance, is among the top countries for biological farming (15.2% of the total) meanwhile being far ahead in implementing AI for the optimisation of irrigation and application of pesticides and herbicides.

In the design industry, AI can take advantage of its capacity to rapidly analyse enormous quantities of data in order to improve the ability to identify the best way to

[62] Stanford Encyclopedia of Philosophy. 2018. *Artificial Intelligence*. https://plato.stanford.edu/entries/artificial-intelligence; Finn (2017); Montagnani (2020); Abriani and Schneider (2020), p. 1326.

[63] AI is defined as the set of methods that allow information to be extracted from large amounts of data and are used to perform a variety of operations, including those of a predictive nature.

[64] Abriani and Schneider (2020); Council of Europe. n.d. What's AI? https://www.coe.int/en/web/artificial-intelligence/what-is-ai.

[65] Comandé (2019), pp. 169–188.

[66] Montagnani (2020), p. 1010.

[67] In essence, AI systems are characterized by their ability to learn autonomously, gradually modifying themselves based on the information they take in and in relation to how they process it.

[68] Abriani and Schneider (2020), p. 1329.

[69] Montagnani (2020), p. 1006.

[70] Mosco (2019), pp. 247–260.

design products based on criteria of circularity, considering both the use of raw materials in production and the object's entire life cycle.

6 The "Sustainable Business Model" for A New Notion of Enterprise

The exercise of the business activity in the three-dimensional space defined by the circular economy gives a new meaning to the creation of value that consequently determines the evolution of the Business Model (BM) in Sustainable Business Model (SBM).[71]

BM can be described as the organizational, management and operational system by which activities that transform resources into results are carried out. It defines, therefore, the way in which the company transforms acquired values by creating a new value. Clearly, the concept of value is central to the design of the business model, both in terms of counting what is introduced into the production system, i.e., the values acquired, and what is produced by the system, i.e., the values released. A recent study aimed at defining a method for measuring circularity starts from the consideration that for an effective measurement it is necessary to consider three aspects, namely the quantity and characteristics of the resources withdrawn (material, energy, and water), used and reintroduced into the system; the environmental impact of the resources used and reintroduced into the system; the economic value of the resources used, and the value when they are reintroduced into the system.[72]

According to the logic of the circular economy, when evaluating the process of economic growth and assessing value creation, it is necessary to take into account not only the economic variable but also the environmental and social variable. From this perspective, the view of neo-classical economics, that a company creates value when the revenues from its activities exceed the costs of carrying out those activities, is obsolete. From the perspective of the bioeconomy, the value creation process must be based on the fact that there are "impassable limits that require the continuous reuse of goods".[73] The definition of value creation must also take account of the fact that there is not only a financial component, but also an economic component in the general sense, as identified above, as well as social and environmental components. This conception of value should be matched by business models "aimed at the pursuit of long-term corporate value creation processes [...] calibrated so as to maximise the various types of capital".[74] In other words, a system allows members

---

[71] CSR Manager Network. 2020. *Creazione di Valore e Sustainable Business Model approccio strategico alla sostenibilità*. https://www.csrmanagernetwork.it/files/DOCUMENTI/CRF_Creazione_di_valore_e_SBM_191020.pdf; Bifulco and D'Aloia (2008).

[72] Georgescu-Roegen (1998); Georgescu-Roegen (2003); Boulding (1981); Costanza (1991).

[73] Georgescu-Roegen (2003).

[74] CSR Manager Network (2020), p. 7.

and stakeholders to verify the increase in value in accordance with different perspectives.

We are still at the early stage of evaluating the negative externalities generated by business activity. The task is to verify the wealth actually produced net of negative externalities, of the social costs produced, then to internalise negative externalities, either voluntarily, as the Corporate Social Responsibility (CSR) claims, or compulsorily, through regulation: "Where the law fails to require corporations to take their externalities into account, corporations rarely take account of their social costs voluntarily".[75] The removal of externalities is, however, extremely far away in all economic systems, despite the increased sensitivity of consumers and businesses and the recognition in some legal systems—such as the English (Companies Act of 2006) and the French (Pacte Law of 2019), just examples, at the European level, of accomplished regulation on the role and the obligation of company directors to balance profit maximization with the interests of stakeholders,—of precise obligations for directors to consider the interests of stakeholders in their management choices. Indeed, Article 1833(2) of the French Civil Code provides that "The company shall be managed in its social interest, taking into account the social and environmental challenges of its activity". Section 172 of the UK Companies Act, entitled "Duty to promote the success of the company", provides that:

(1) A director of a company must act in the way he/she considers, in good faith, would be most likely to promote the success of the company for the benefit of its members as a whole, and in doing so have regard (amongst other matters) to—

  (a) the likely consequences of any decision in **the long term,**
  (b) the interests of **the company's employees,**
  (c) the need to foster the company's business relationships with **suppliers, customers and others,**
  (d) the impact of the company's operations on the **community and the environment,**
  (e) the desirability of the company maintaining a reputation for high standards of business conduct, and
  (f) the need to act fairly as between members of the company.

In an even more avant-garde perspective, it is important to overcome the logic that ESG factors (Environment, Social, Governance) are considered only as risk factors with respect to business activities, and not as interests on the basis of which business activities should be managed. The management of ESG risks[76] and the relative transparency aim,[77] instead, is to trigger a virtuous circuit, i.e., to push companies to consider such factors "to the extent that this will be appreciated or, conversely, not

[75] Johnston (2011), pp. 221–224.
[76] Pellegrini et al. (2020).
[77] Fortunato (2019), pp. 415–428.

rewarded by the market, i.e., by current and potential investors".[78] Therefore, the criterion for making information public "is its relevance to understanding the impact on the business activity,[79] i.e., how non-financial issues affect the main risks and how they are managed.[80]

7 Business Models

In a broader perspective, "circular growth" business models have been proposed,[81] but each one only pursues individual aspects of the circular economy.

The first of these models is identified as the 'circular supply chain' and requires that the energy and raw materials needed to make products be fully renewable, recyclable or biodegradable, replacing linear materials that are derived from virgin resources, are not recyclable and are frequently toxic and polluting.[82] The model does not simply aim to minimise bad materials, but to produce something that has value, in the sense of the circular economy, and therefore makes economic sense.[83] This means organising the business in such a way as to "do things right from the start".[84]

The second BM is the so-called "recovery and recycling" model according to which the waste, according to the linear economy model, "is reintroduced for other uses, effectively eliminating not only waste but the concept of waste itself".[85] In this way, the wealth produced by the business derives not only from the final product but also from any by-products.[86]

The third model is called "product life extension". According to the linear economy model, companies have organised their strategy with the objective of increasing product volume, following the logic that revenue increase depends solely on the increase in the number of units of product produced and sold: "The Product Life Extension business model extends the life cycle by generating revenue through longevity rather than volume".[87]

---

[78] Bruno (2018), p. 980.

[79] Ibid. 984.

[80] Ibid. 985.

[81] Lacy et al. (2016); Dallocchio, Maurizio, Leonella Gori and Emanuele Teti. n.d. I *Modelli di business dell'economia circolare*. https://rivista.microcredito.gov.it/opinioni/archivio-opinioni/763-i-modelli-di-business-dell%E2%80%99economia-circolare.html.

[82] Lacy et al. (2016); Dallocchio et al. (n.d.).

[83] Lacy et al. (2016), p. 63.

[84] Ibid. 85.

[85] Ibid.

[86] Ibid.

[87] Ibid. 107–108.

The fourth BM is the so-called "sharing platform" which connects, via a special platform, the owners of products with those interested in using them.[88] Use of the platform increases productivity by not leaving products unused and, at the same time, reduces the demand for new goods by allowing more customers to use the same resources.[89]

The fifth BM for circular growth is the "product as a service" model, whereby ownership of the product remains with the producer who offers it to one or more customers as a service, including everything necessary to keep it in use.[90] The customer is thus transformed from a consumer of the product into a user of the service.[91]

However, although the models summarized above are compatible with the principles of the circular economy, they do not make the logical leap that enables the company to live in the three-dimensional space designed by the circular economy.

The problem posed by the circular economy is quite different, since it is a matter of designing and managing business activities with equal consideration for the economic, social and environmental aspects, detecting the demands that come from each of these poles and carrying out business activities in such a way as to coordinate and satisfy all the demands thus identified. The BM, structured in this way, evolves into the Sustainable Business Model (SBM), capable of defining the company's objectives "also in social and environmental terms and not only from an economic and financial point of view".[92]

The adoption of an SBM, however, requires more research and management of a huge amount of data. This task would be difficult to solve with the use of traditional techniques and, in any case, the adoption of such a model would imply a reconsideration of the value to be attributed to the adequacy of corporate structures. IA is the tool that can offer a solution to this need, since it has already been considered crucial in the management of ESG factors,[93] due to its ability to process an enormous quantity of data and identify important correlations between them,[94] as well as its ability to adapt to the environment in which it operates.

The efforts required by companies, in terms of the amount of data to be acquired and processed, in order to operate in the dimension of the circular economy are, however, more intense than those required for the simple management of ESG risk factors. This factor makes recourse to AI even more useful and, perhaps, indispensable.

---

[88]Ibid. 131.

[89]Ibid.

[90]Ibid. 153.

[91]Ibid. 154.

[92]CSR Manager Network (2020), p. 29.

[93]Montagnani (2020), p. 1006.

[94]Montagnani (2020), p. 1003.

8 Conclusion

The design and management of business activity which takes equal consideration of economic, social and environmental aspects, as required by circular economy, shows the inadequacy of the principles and rules that inspire the corporate purpose from the traditional organizational, technical, administrative and accounting point of view, centered essentially around the interests of shareholders and creditors, from a predominantly economic-financial perspective. This standpoint questions the essence of the company—an entity which is no longer based on the notion of capital or on the stakeholders—"groups and individuals who benefit from or are harmed by, and whose rights are violated or respected by, corporate actions"[95] (Evan and Freeman 1988, p. 100).

References

Abriani N (2022) Le categorie della moderna cibernetica societaria tra algoritmi e androritmi: "fine" della società e "fini" degli strumenti tecnologici. Giurisprudenza commerciale 1:761

Abriani N, Schneider G (2020) Il diritto societario incontra il diritto dell'informazione. IT, Corporate governance e corporate social responsibility. Rivista delle società 5/6:1326

Accenture (2017) Boost Your AIQ Transfroming into an AI Business. https://newsroom.accenture.com/news/artificial-intelligence-leaders-generate-greater-shareholder-value-finds-new-report-from-accenture-research.htm

Bifulco R, D'Aloia A (eds) (2008) Un diritto per il futuro. Teorie e modelli dello sviluppo sostenibile e della responsabilità intergenerazionale. Jovene, Napoli

Boulding K (1981) Evolutionary economics. Beverly Hills

Bovino C (2014) Verso un'economia circolare: la revisione delle direttive sui rifiuti. Ambiente & sviluppo: consulenza e pratica per l'impresa e gli enti locali 10:682–691

Bruno S (2018) Dichiarazione "non finanziaria" e obblighi degli amministratori. Rivista delle società 4:974–1020

Cineca (n.d.). https://www.cineca.it/temi-caldi/Leonardo

Cocconi M (2020) La regolazione dell'economia circolare. Sostenibilità e nuovi paradigmi di sviluppo. FrancoAngeli

Comandé G (2019) Intelligenza artificiale e responsabilità tra "liability" e "accountability". Il carattere trasformativo dell'IA e il problema della responsabilità. Analisi Giuridica dell'Economia 1:169–188

Costanza R (ed) (1991) Ecological economics: the science and management of sustainability. Columbia University Press

Council of Europe (n.d.) What's AI? https://www.coe.int/en/web/artificial-intelligence/what-is-ai

CSR Manager Network (2020) Creazione di Valore e Sustainable Business Model approccio strategico alla sostenibilità. https://www.csrmanagernetwork.it/files/DOCUMENTI/CRF_Creazione_di_valore_e_SBM_191020.pdf

Dallocchio M, Gori L, Teti E (n.d.) I Modelli di business dell'economia circolare. https://rivista.microcredito.gov.it/opinioni/archivio-opinioni/763-i-modelli-di-business-dell%E2%80%99economia-circolare.html

[95] Evan and Freeman (1988).

De Leonardis F (2017) Economia circolare: saggio sui suoi diversi aspetti giuridici. Verso uno Stato circolare. Diritto amministrativo 25(1):163–207

De Leonardis F (ed) (2019) Studi in tema di economia circolare. EUM-Edizioni Università di Macerata

De Leonardis F (2020) Il diritto dell'economia circolare e l'art. 41 della Costituzione. Rivista Quadrimestrale di Diritto dell'Ambiente 1:50–72

Ellen MacArthur Foundation (2019) Completing the picture. How the circular economy tackles climate change. https://emf.thirdlight.com/link/dcijanpohgkd-oblthh/@/preview/5

Enriques L, Zetzsche DA (2020) Corporate technologies and the Tech Nirvana Fallacy. Hast Law J 72(1):55–98

European Commission (2011) A renewed EU Strategy 2011-14 on Corporate Social Responsibility. https://eur-lex.europa.eu/LexUriServ/LexUriServ.do?uri=COM:2011:0681:FIN:EN:PDF

European Commission (2014) The Circular Economy. Connecting, Generating and Retaining Value. https://op.europa.eu/en/publication-detail/-/publication/c8cfd1ae-6285-40ba-879f-f2e78e4c2b6e

European Commission (2020a) A new Circular Economy Action Plan. https://eur-lex.europa.eu/legal-content/EN/TXT/?qid=1583933814386&uri=COM:2020:98:FIN

European Commission (2020b) White paper on artificial intelligence - A European approach to excellence and trust. https://ec.europa.eu/info/publications/white-paper-artificial-intelligence-european-approach-excellence-and-trust_en

Evan W, Freeman ER (1988) A stakeholder theory of the modern corporation: Kantian capitalism. In: Beauchamp TL, Bowie NE (eds) Ethical theory and business. Prentice Hall, Englewood Cliffs, N.J

Federal Office for Spatial Development ARE. Brundtland Report 1987. https://www.are.admin.ch/are/it/home/sviluppo-sostenibile/cooperazione-internazionale/agenda2030/onu-_-le-pietre-miliari-dello-sviluppo-sostenibile/1987%2D%2Drapporto-brundtland.html

Ferrara R (2018) Brown economy, green economy, blue economy: l'economia circolare e il diritto dell'ambiente. Diritto e processo amministrativo 12(3):801–835

Finn (ed) (2017) What algorithms want, imagination in the age of computing. MIT Press

Fortunato S (2019) L'informazione non-finanziaria nell'impresa socialmente responsabile. Giurisprudenza commerciale 46(3):415–428

Fraccia F (2010a) Sviluppo sostenibile e diritti delle generazioni future. Rivista Quadrimestrale di Diritto dell'Ambiente 0:15

Fraccia F (2010b) Lo sviluppo sostenibile. La voce flebile dell'altro fra protezione dell'ambiente e tutela della specie umana. Editoriale Scientifica, Napoli

Fraccia F (2012) Il principio dello sviluppo sostenibile. In: Renna M, Saitta F (eds) Studi sui principi di diritto amministrativo. Giuffrè, Milano

Freeman ER, Harrison JS, Zyglidopoulos S (2018) Stakeholder theory: concepts and strategies. Cambridge University Press, Cambridge

Georgescu-Roegen N (2003) Bioeconomia Verso un'altra economia ecologicamente e socialmente sostenibile. Mauro Bonaiuti (ed) Bollati Boringhieri

Georgescu-Roegen N (1998) Bioeconomia. Verso un'altra economia ecologicamente e socialmente sostenibile. Bollati Boringhieri

Goodman J, Salleh A (2013) The 'Green Economy': Class Hegemony and Counter-Hegemony. Globalizations 10(3):411–424. https://doi.org/10.1080/14747731.2013.787770

Greco P, Salimbeni AP (2003) Lo sviluppo sostenibile: dal vertice di Rio a quello di Johannesburg. B. Mondadaori, Milano

Johnston A (2011) Facing up to social cost the real meaning of corporate social responsibility. Griffith Law Rev 20(1):221–224

Joshi N (2019) How IoT and AI Can Enable Environmental Sustainability. Forbes, 4 September 2019

Kaplan J (2018) Intelligenza artificiale Guida al futuro prossimo. Luiss Università Press

Keay AR (2011) The corporate objective. Edward Elgar Publishing

Lacy P, Lamonica B, Rutqvist J (2016) Circular Economy Dallo spreco al valore. Egea
Lanza A (2002) Lo sviluppo sostenibile. Il Mulino, Bologna
Lomonaco R (2004) Sviluppo sostenibile e difesa dei diritti umani, Roma
Mattei U (2011) Beni comuni. Un manifesto. Laterza, Roma-Bari
Millon D (2013) Radical shareholder primacy. Un St Thomas Law J 10(4):1013–1044
Moliterni A (2020) La sfida ambientale e il ruolo dei pubblici poteri in campo ambientale. Rivista Quadrimestrale di Diritto dell'Ambiente 2:32–71
Montagnani ML (2020) Intelligenza artificiale e *governance* della "nuova" grande impresa azionaria: potenzialità e questioni endoconsiliari. Rivista delle società 65(4):1003
Mosco GD (2019) Roboboard. L'intelligenza artificiale nei consigli di amministrazione. Analisi Giuridica dell'Economia 1:247–260
Munda G (1997) Economia ambientale, economia ecologica e concetto di sviluppo sostenibile. Valori ambientali 6(2):213–233
Pauli G (2015) Blue economy 2.0. Edizioni Ambiente, Milano
Pellegrini CB, Dallocchio M, Parazzini E (2020) Valutazione d'azienda nel mondo ESG. Egea, Milano
Pepe V (2002a) Lo sviluppo sostenibile tra diritto internazionale e diritto interno. Rivista giuridica dell'ambiente 17(2):209–243
Pepe V (2002b) Lo sviluppo sostenibile tra governo dell'economia e profili costituzionali. La Tribuna, Piacenza
Scotti E (2019) Poteri pubblici, sviluppo sostenibile ed economia circolare. Il diritto dell'economia 1:493–529
Sjåfjell B, Bruner CM (2020) Corporate law, corporate governance e sustainability. Cambridge University Press, Cambridge
Stanford Encyclopedia of Philosophy (2018) Artificial Intelligence. https://plato.stanford.edu/entries/artificial-intelligence
Stout LA (2012) The shareholder value myth. Berret-Koehler Publishers, San Francisco
Strine LE Jr (2019) Toward Fair and Sustainable Capitalism. SSRN. https://ssrn.com/abstract=3461924
UNDESA (Division for sustainable development), A Guidebook to Green Economy, 2012.
United Nation (2002) The Johannesburg Declaration on Sustainable Development. From our origins to the future. https://www.un.org/esa/sustdev/documents/WSSD_POI_PD/English/POI_PD.htm
United Nation, The further we want 2012., https://www.minambiente.it/sites/default/files/archivio/allegati/rio_20/the_future_we_want_english.pdf
United Nation, The Paris Agreement 2015., https://unfccc.int/process-and-meetings/the-paris-agreement/the-paris-agreement
United Nation., The Stockholm Declaration 1972, https://www.un.org/en/conferences/environment/stockholm1972
United Nation., The Rio Declaration on Development and the Environment 1992, https://www.un.org/en/development/desa/population/migration/generalassembly/docs/globalcompact/A_CONF.151_26_Vol.I_Declaration.pdf
Vaioa D, Assunta RP, Hassanb R, Escobar O (2020) Artificial intelligence and business models in the sustainable development goals perspective: a systematic literature review. J Bus Res 121:283–314. https://doi.org/10.1016/j.jbusres.2020.08.019
Vinuesa R, Azizpour H, Leite I, Balaam M, Dignum V, Domisch S, Felländer A, Langhans SD, Tegmark M, Nerini FF (2020) The role of artificial intelligence in achieving the Sustainable Development Goals. Nat Commun 11:233. https://doi.org/10.1038/s41467-019-14108-y
Webster K (2017) The circular economy. A wealth of flows. Ellen MacArthur Foundation Publishing

Part II
Blockchain for Sustainable Development: The Potential of Tokenization

Token, Tokenization and Sustainable Development

Raffaele Lener and Salvatore Luciano Furnari

Abstract Financial innovation has given great importance to sustainability. The Sustainable Finance Strategy represents one of the central points of the European agenda, and Fintech offers important opportunities in this field. However, some "valuable" Fintech instruments seem to not have been adequately considered. The reference is to Bitcoin and to the "crypto-industry". The most common public belief about Bitcoin and sustainability is that Bitcoin is polluting the ecosystem. Immediately, the concept is extended to Distributed Ledger Technologies (DLTs), to blockchain and to all related innovations. Notwithstanding this, some research shows that there is still some uncertainty on the precise amount of energy used by Bitcoin-related activities and on how to calculate it. The scope of this paper is to give some clarity on this uncertainly in order to show, on the one hand that DLT is not "polluting", but that pollutions come from the way each miner decides to conduct their businesses. On the other hand, a DLT system can be considered sustainable thanks to its capacity to solve a various number of environmental problems "related" to how businesses are conducted.

Keywords Bitcoin · Blockchain · DeFi · Sustainable development

The article was previously published in the journal Law&Digital Technologies, 2022, 2(1): 16–23.

R. Lener (✉) · S. L. Furnari
University of Rome Tor Vergata, Roma, Italy
e-mail: raffaele.lener@leplex.it; salvatore.furnari@leplex.it

L. V. Sannikova (ed.), *Digital Technologies and Distributed Registries for Sustainable Development*, Law, Governance and Technology Series 64,
https://doi.org/10.1007/978-3-031-51067-0_3

1 Introduction

Financial innovation has given great importance to sustainability. The Sustainable Finance Strategy represents one of the central points of the European agenda, and Fintech offers important opportunities in this field.[1] However, some "valuable" Fintech instruments seem to not have been adequately considered. The reference is to Bitcoin and to the "crypto-industry". In January 2023 , Bitcoin had a market cap of $438 billion, being at the 17th place in the ranking of assets market cap, immediately after Tesla ($439 billion) and before Meta (Facebook) ($371 billion).[2] In addition, the total market cap of the first 10 crypto-assets was equal to $856 billion.[3]

The most common public belief about Bitcoin and sustainability is that Bitcoin is polluting the ecosystem.[4] Immediately, the concept is extended to Distributed Ledger Technologies (DLTs), to blockchain and to all the related innovations.

Research on Bitcoin sustainability started in 2015.[5] The controversial question of Bitcoin sustainability arises from the fact that Bitcoin protocol has been programmed to let this cryptocurrency become a scarce resource. The more Bitcoins are *mined*, the harder it is to mine more. This leads to the increased amount of energy spent in the process of mining.[6]

There is a common conception that Bitcoin creation is an energy-consumptive activity. Cambridge University has conducted a comparison study on Bitcoin energy consumption.[7]

The research showed that Bitcoin ranks 34th in annual electricity consumption at 102.8 TWh, which is more than the electricity consumption of the entire country of Philippines (at 98.5 TWh), but less than of Netherlands (at 113.3 TWh per year).[8] However, Bitcoin's annual electricity consumption is still far less than that of Gold mining (131 TWh per year).

Notwithstanding the above, some research shows that there is still some uncertainty on the precise amount of energy used by Bitcoin-related activities and on ways of energy consumption calculation.[9]

The aim of this paper is to clarify this uncertainty in order to show, on the one hand, that DLT is not "polluting", but that pollutions come from the way each miner decides to conduct their business. On the other hand, a DLT system can be

[1] Macchiavello and Siri (2020).

[2] CompaniesMarketCap.com. January 23 2023. https://companiesmarketcap.com/assets-by-market-cap/.

[3] CoinGecko. January 232,023. https://www.coingecko.com/.

[4] Business Standard (2019), McMaster (2017), Rowlatt (2021) and Aratani (2021).

[5] Giungato et al. (2017), pp. 2214.

[6] Ibid.

[7] Cambridge Centre for Alternative Finance. n.d. https://ccaf.io/cbeci/index.

[8] However, the first country is China (6453.17 TWh per year) followed by USA (3843 TWh per year) and India (1277 TWh per year).

[9] Stoll et al. (2019), pp. 1647–1661.

considered sustainable thanks to its capacity to solve a various number of environmental problems "related" to how businesses are conducted.

2 Functioning of Blockchain Technology. Where Does the Danger for the Ecosystem Come From?

2.1 Blockchain and DLT. A Brief Overview

Blockchain is a form of DLT, a technology based on a decentralized register managed by a peer-to-peer network. This database can keep the record of the transactions made by its participants independently of a centralized authority that manages the system.

DLT allows transparency and full disintermediation.

Transparency is derived from the fact that most blockchain protocols can be consulted by anyone. This means that anyone can read all the transactions made, as well as the number of crypto-assets contained in a wallet or the code of the software (i.e., smart contracts) deployed on it.

Disintermediation means, *inter alia*, that anyone can participate in the network to run a "node". The nodes possess a full copy of the register containing the information on all the transactions made and on all the software deployed within the blockchain. Nodes, also, participate in block-validation activities, checking if the information which can be stored in the next block is coherent with the previously recorded one.[10]

2.2 Energy Consumption and Consensus Mechanisms

A distributed ledger requires a secure mechanism to store information. Each node – owning a copy of the register—could *easily* change the data stored, creating potential confusion or, at the worst, cheating. To store information in a secure way, in order to avoid double spending and manipulations, blockchain adopts specific systems known as "consensus mechanisms". According to a common definition a consensus mechanism is "a fault-tolerant mechanism that is used in computer and blockchain systems to achieve the necessary agreement on a single data value or a single state of the network among distributed processes or multi-agent systems, such as with

[10]For instance, if in block no. 8 Alice is recorded to possess 10 tokens, and if she decides to transfer 5 tokens in the following block, the information that needs to be stored is that Alice holds 5 tokens. If a node "writes down" that Alice still holds 10 tokens, that information is not stored because the block is not validated.

cryptocurrencies".[11] One of the first adopted (and actually used by Bitcoin protocol) is the one called Proof-of-Work (PoW).

A PoW-based blockchain requires validators to have done "*some works*" to connect blocks of information. For instance, Bitcoin protocol requires to solve a sort of "*cryptographic puzzle*" to find a number with specific characteristics that permits the connection of two blocks of information. Here, the first miner who solves the puzzle receives a payment from the protocol, equal to a specific amount of Bitcoin.

Thanks to this system, each node of the blockchain can verify if the data of the copy of the register has been altered. An alteration, indeed, will imply that each block shall be connected by a number that is different from the one recorded and already shared with the network.

In order to benefit from its protection, PoW requires miners to spend lots of energy. This energy is consumed by the powerful processors used by miners to solve the "cryptographic puzzle" before other miners get *per se* the rewards. The PoW system makes altering the information stored a very difficult activity.

Indeed, in order to alter a sequence of blocks, the PoW system requires the solving of the "cryptographic puzzle" for each block. In order to do so, it is necessary to spend all the energy that has already been spent to "solve the puzzle" of all the blocks that a "malicious" node wants to alter. In addition, this energy consumption activity has to be completed before an additional block is created.[12]

2.3 Token and Tokenization

The information that could be recorded on a DLT can have the form of (1) tokens and/or (2) smart contracts.

A token can be defined as a record in favor of a participant that lets them be recognized by the entity who released the token as the holder of a precise amount and/or kind of rights. So, from a *technical* point of view, a token is nothing else than a simple registration in favor of its holder. From a *functional* point of view, it is the informatic "instrument" through which the holder may exercise a precise kind of rights towards the offering entity.[13] Those rights are, indeed, the subject of the offer itself, that is, what an investor will gain in buying the offered token.[14] Sometimes they confer access to a service provided by the platform (utility tokens). In other cases, they confer voting or economic rights towards the issuing company (investment tokens).

[11] Investopedia.com (n.d.).

[12] This, according to Bitcoin protocol, happens each 10 min; 13 s is the time to create a new block on the Ethereum blockchain.

[13] Furnari (2018).

[14] Ibid., 144.

After being issued, tokens can be sent to or exchanged with other participants. They are easily stored by their owners using a "wallet", that is a software that stores the private key to interact with the blockchain. Wallet services can be offered by third parties, but more often token-holders uses non-third-party wallets. This enables each token holder to maintain a complete power on their tokens. As per the mere holding, the exchange of tokens does not require any third party to be completed. These two characteristics highlight how, not only the infrastructure (i.e., the DLT), but also the "objects" of the infrastructure (i.e., the tokens) are disintermediated.

This high level of disintermediation gives the rise to a phenomenon called *tokenization*. Tokenization is the possibility to turn a specific "right" into a token. The only existing limits are, at most, the legal ones.[15] Tokenized rights can circulate and be held without the limits given by potential gatekeepers (i.e., intermediaries) or technological limitations.[16]

Advantages given by tokenization are various. One of the most important is the ability to transfer the right incorporated in them by just transferring the token.

2.4 *Origins of Tokens*

Tokens are usually created by a blockchain protocol or by a smart contract. Smart contracts are algorithmic sequences (i.e., softwares), elaborated by *virtual machines*, sorts of "big *phantom* computers" created by the power of calculation given by the nodes of the blockchain.

As all information is recorded on the blockchain, smart contracts acquire the same fundamental qualities of the blockchain on which they "run". This fact makes them suitable for the execution of contracts (from which they took their name). Indeed, as on all software, smart contracts are self-executing; but being launched on a blockchain, they are also unstoppable. If a smart contract is programmed to perform a determined action, it will work until the action is completed. If a precise mechanism to stop its functioning has not been "programmed" by the party who launched it, nobody can stop its functioning without taking control of 51% of the power of calculation alimenting the blockchain.[17]

This also means that a smart contract completely lacks the human interaction for its execution. In this way, it can be used to perform obligations deriving from a real contract that could be written within the smart contract itself.[18] A contract of this

[15] The rights that most commonly are tokenized are credit rights under various forms. Utility tokens and investment tokens are the tokenized version of consumer (the right to obtain a good or a service) or investment (the right to vote or to dividends) rights.

[16] Lener and Furnari (2020), pp. 33–52.

[17] Furnari (2019).

[18] To be more precise, they can perform the role of an "online vending machines" to highlight their basic function of executing a predetermined action in response of a precise input.

kind could help the managing of the performance execution since there is no need for the interpretation of the contract terms. The parties to the agreement do not need to trust each other before the conclusion of the agreement since its execution is fully automated. This principle applies particularly to the collection of money through the launch of an Initial Coin Offering (ICO) or to providing different kinds of services (including financial ones). If the collection of money is managed using a smart contract, this program will automatically deliver the token in exchange for the money received.

Finally, smart contracts can be also used by the issuer to "strongly" grant the right attached to the token distributed. For instance, if a token grants access to a specific service of the issuer, and if the access is regulated with the use of a smart contract, the buyer of the token could be sure that they will enjoy the service they paid for.

3 What Is Sustainability

Much research has provided different definitions of *sustainability*.[19] However, there is still the lack of a standard one. The lack of a standard definition of sustainability arises from the fact that the topic is being researched in various fields: from ecology to corporate governance and business development.[20]

Notwithstanding this, it is still possible to agree on the fact that sustainability is a quality of a system that permits it to (1) reduce or eliminates various kind of costs; (2) avoid or resolve scarcity problems; and (3) avoid or react to various type of fragility. The more a system is "sustainable", the more it manages the solution of those difficulties (costs, problems, and fragilities).[21]

The adoption of this definition to blockchain-based systems leads to the conclusion that it is not true to say that blockchain technology is unsustainable.

At most, what is truly unsustainable is the way some miners conduct their job. Here it is clear that, if the energy used to perform PoW activities comes from carbon fuel, the emissions cause pollutions that are an important *cost* for the environment, creating (instead of resolving) various kinds of *fragilities*. This clarification is fundamental since it demonstrates that blockchain is not polluting *per se*, but its environmental impact depends on how a miner decides to power their engines. Indeed, it is clear that the same environmental problems will not arise if the same activities are conducted using solar or similar energies and more eco-friendly sources.

However, it is still possible to argue that, as far as a system requires great amounts of energy to operate, it could not be called fully "sustainable". This implies the need

[19] Allen and Hoekstra (1993), pp. 98–107; Moore et al. (2017), p. 110.

[20] Ferrarini (2020); De Hoo and Olaerts (2011).

[21] Di Paola (2016).

to take into account some additional considerations in order to show how a blockchain-based system (and so tokenization) can be seen as sustainable.

4 Sustainable By Technology

One of the first attempts to solve blockchain environmental issues deals with the substitution of PoW consensus mechanism with more energy efficient solutions. One of these is Proof-of-Stake (PoS).[22]

The PoS consensus mechanism overcomes the problem of selecting the validator nodes on the ground of the amount of work performed (and so on the energy consumed). Here validators are selected randomly between those who possess a certain amount of tokens (the "stake") and some additional related characteristics.[23]

For instance, in the "*Randomized Block Selection*" method, validators are chosen randomly among those who have a combination of the lowest hash value and the highest stake. In the "*Coin Age Selection*" method, validators are chosen based on how long their tokens have been staked.[24] Here once a node has validated a block, he/she must wait a certain period of time before being able to validate another block. In this way, it is possible to prevent bigger validators to obtain the control over the blockchain.

PoS mechanism demonstrates the existence of sustainable ways to make a blockchain work. Indeed, as distinct from PoW, PoS is not based on the energy consumed to permit the functioning of the validation process.

Its sustainability, in addition, does not alter its safety from the cyber secure point of view; PoS systems are still secure as PoW.[25]

This is true thanks to the fact that the staked tokens function as a "bail" for the validator to not indulge in fraudulent activities. Indeed, fraudulent activities are sanctioned detaining part of the token staked.

The other aspect that increases security in PoS-adopting blockchains is the higher level of decentralization that is possible to achieve. In comparison with PoW systems—where the amount of power that is necessary to validate a block induces the creation of big mining pool, with a clear oligopoly risk—PoS makes it possible for almost everyone to run a node because it lowers the entry barriers of validations activities. This fact favours a wider distribution of the tokens needed to run the node, making attacks more difficult to be performed. This is truer because if in PoW-adopting blockchains, to make a so-called "51% attack" there is the need to

[22] Saleh (2021), pp. 1156–1190; Chohan (2018); He et al. (2020).

[23] Chohan (2018).

[24] Token age is calculated by multiplying the days the token has been held as stake by the number of tokens staked.

[25] Thapa et al. (2019).

control the 51% of the mining power, in PoS it is necessary to possess the 51% of all the tokens issued.[26]

5 Sustainable By Business

Apart from technological innovations, a blockchain based system could be considered sustainable thanks to the innovative types of business models, based on tokenization, that they permit to develop.

The use of tokenization can change the way in which business~~es~~ is done, making it more sustainable. Wide research exists, for instance, in the field of food and, more in general, on the supply chain. In this sector, the advantages for sustainability in using blockchain systems have already been proved[27] (Lund et al. 2019; Sulkowski 2018; Medhi 2019).

Another sector studied from this point of view is the financial one. This industry can be revolutionized in two different ways. On the one hand, blockchain technology can implement the way in which financial services ~~are~~ have *already* been provided to the public. On the other, blockchain technologies can foster *new ways* to provide financial services.

The first of these two sides of the same coin has been already extensively analysed.[28]

With sustainability defined as "the quality of a system to reduce or eliminate various kinds of costs", a technological innovation would be more sustainable, with increased potential capacity to change completely the way a business is conducted.

From this point of view, it should be noted that tokenization allows the offering ~~of~~ financial services without the need for *financial intermediaries.* This is possible in what is called "Decentralized Finance" (or DeFi), a new way in which financial services may be offered by algorithmic protocols.

DeFi is a very recent evolution of finance, first appearing in 2017 on the Ethereum Blockchain. It started with the MakerDAO Protocol[29] and rapidly evolved into a rich ecosystem, whose value has reached nearly $87 billion in May 2021.

Taking inspiration from Bitcoin and its underlying technology, DeFi permits financial services to be offered without the need ~~of~~ for intermediaries and central authorities to manage and control the system. In this way, protocol-base intermediaries (as opposed to the human-based ones) can reduce the costs for a single transaction while offering users a direct control over their financial assets.

[26] John et al. (2020).

[27] Lund et al. (2019), pp. 16–23; Sulkowski (2018), pp. 303–345; Medhi (2019).

[28] Peters and Panayi (2015); Harris and Wonglimpiyarat (2019), pp. 625–639; Guo and Liang (2016); Popova and Butakova (2019), pp. 1764–1768; Wu and Liang (2017), pp. 219–224; Cocco et al. (2017), pp. 25; Lu et al. (2019), pp. 1008–1014; Wang et al. (2019), pp. 1127–1140.

[29] For more information, please see the official website https://makerdao.com/en/.

So, DeFi substitutes traditional financial intermediaries deploying unique business models which function using protocols running on blockchain technology. A protocol can be defined as an algorithm or as a software embodied in a smart contract. Protocols offer their services independently from a physical or a legal person, directed by a group of developers through a decentralized organization or even a Decentralized Autonomous Organization.

This circumstance reduces the number of people and the related costs (also for the environment) traditionally correlated with the providing of financial services.

A brief overview of the most common protocols of DeFi ecosystem ~~is~~ will be useful to understand how it is possible to offer services very similar to traditional financial services, through a series of smart contracts deployed on a blockchain.

Lending Protocols offer services very similar to the ones offered by banks in lending money. Decentralized Exchanges (DEX) using the Automated Market Makers (AMM) system permits the creation of market infrastructures and the managing of markets for tokens. Launchpad Platforms permits the offering of crypto-assets to the public, resembling traditional IPOs through ICOs.

5.1 Lending Protocols

Lending protocols allow users to become lenders or borrowers of tokens. This becomes possible via the use of smart contracts that permit the platform user to "deposit" tokens within a smart contract to receive back (automatically) an interest on the assets deposited.[30]

The tokens deposited become available for borrowing to other users. Who wants to borrow tokens and provide a guarantee that at the time of the borrowing, they will be valued more than the borrowed tokens? In particular, for each asset lent is associated with a *collateralization ratio* which is the limit a borrow can be obtained providing the asset as a guarantee. This means that, initially, all borrows are over-collateralized. A drop in the value of the collateralized tokens may cause the collateralization ratio limit to be exceeded. If this happens, the protocol sells the assets given as a guarantee, repays the loan, and gives back the difference minus a penalty (that usually is the 10% of the value liquidated).

Lending protocols permits users to obtain liquidity while not selling their token holdings. In this way, they may still benefit from potential increases in value. Another benefit of using a Lending protocol could be simply the intention to increase their leverage on certain trading positions. Finally, the Lending Protocol permits the short-selling[31] of assets.

---

[30] Castro-Iragorri et al. (2021).

[31] To do so, it is necessary to borrow the asset that the user wants to short sell, changing it on a DEX for a stable coin and then repaying back the borrow buying back the borrowed token when their value just decreased.

5.2 DEXes

DEXes are DeFi protocols that let users exchange tokens for other tokens. The most famous DEXes, instead of running a "classic" order book, implement AMM systems to increase their productivity and to avoid lack of liquidity.[32]

Considering their importance in the functioning of the DEX, an AMM could actually be defined as a type of decentralized exchange (DEX) that uses a specific formula to price assets.

An AMM needs the creation of trading pairs to work. Its innovation lies in the fact that there is no need to have an actual counterparty (i.e., another trader) to perform the exchange. Indeed, users interact with a smart contract in which a "pool" of the two assets composing the pair has been created by other users. Those users act as Liquidity providers and receive a fee for all the trades performed using their pooled assets.

When the tokens are deposited, the pool is composed of an amount of assets having the same value. In this way, when an exchange is requested, the protocol exchanges the required asset at the same value as the one provided, less fee, which increases the pool.

If the demand for one of the two assets increases, so does its price. So, the protocol rebalances the amount of tokens composing the pool in such a way that they always have the same value. This mechanism creates a specific risk for liquidity providers called Impermanent Loss.[33]

5.3 Launchpads and ICO

Launchpads are platforms that permit users to participate in Initial Coin Offerings (ICO).

---

[32] Mohan (2020); Capponi and Jia (2021); Aspris et al. (2021), pp. 1–10.

[33] Impermanent Loss can be defined as the loss incurred in deciding to participate in the liquidity pool instead of holding the single asset. Indeed, if the value ratio of the two tokens changes a lot, the protocol distributes part of the increased value of one token to the other one, selling the first and buying the second. This implies a reduction of the amount of the token whose value increases. In this circumstance the liquidity providers would have had a grater gain simply holding the tokens instead of adding funds to a pool.

The loss of the liquidity provider increases when the value of one of the tokens drops significantly. In this case, indeed, the liquidity provider suffers a decrease also of the paired token.

The terms "impermanent" means that the loss does not actually suffer by the liquidity provider until he/she decides to remove the liquidity. Changing of the market value of the two tokens may always recreate the conditions before the token dropping value.

An ICO is an offering to the public of tokens for the collection of money. From this point of view, it can be defined as a more advanced technological version of crowdfunding.[34]

A particular phase of the collecting procedure, which is worth mentioning, is the so-called "Airdrop", a way of spreading new tokens for free. The scope of airdrops is to speed up token adoption with a clear marketing purpose.[35]

An ICO is usually anticipated by the publication of a white paper, a document presenting ICO scope and characteristics. Its content is not pre-established, but it usually deals also with the technical description of the token and of the smart contract involved in the offer.

The participant in an ICO receives in exchange for their participation a token which can be programmed to play a wide range of roles in the functioning of the company. Traditional classifications recognize three main categories: cryptocurrency, utility and investment tokens. But the classification exercise is not always simple, because of the presence of so-called hybrid tokens, tokens which do not fit any of the three traditional categories since they share the characteristics of two or more of them, without being classified as an autonomous category.

The term ICO is generally used as a broader term to refers to the offering of tokens to the market, in exchange for fiat money of other tokens. The DeFi rapidly evolving markets have seen the rise of different ways to conduct this offering.

One of the most common is the Initial Exchange Offering (IEO). IEO can be simply defined as an ICO conducted on a launchpad platforms, traditionally a cryptocurrency exchange.

The first fundamental characteristic of an IEO is the fact that it is *intermediated.* Indeed, ICOs generally are conducted directly on the website of the issuing entity. From the point of view of the promotion of the offer, the launchpad platforms have the same role as a crowdfunding platform. A potential investor within the platform has the possibility of choosing between different "investment" solutions. So launchpads grant important advantages to the issuer, that is, a prepared crowd of potential investors.

The use of an intermediary platform also gives some advantages to potential investors who may trust the fact that the exchange has performed a due diligence on the token offering, in order to avoid fraud or scum offering. Due diligence is usually conducted in the ~~first~~ interests of the platform in order to avoid damages to its image.

[34] Furnari (2021), pp. 101–117.

[35] This could be possible because additional token creation is free of direct costs for the entrepreneur. Indirect costs are, instead, related to the value dilution of injecting new liquidity into a system.

6 Conclusion

It is immediately obvious that DeFi can be seen as more sustainable than a traditional financial system. This actually can be considered true due to the fact that "DeFi algorithms" do not need buildings and energy consuming computers or people who need to commute to work by cars, trains or aeroplanes. Protocols consume only the energy of the blockchain, and new blockchains using PoS instead of PoW are more sustainable.

However, as of today, the traditional financial system will not disappear. This can be considered true for, at least, two main reasons. First, the legal framework related to this new technology is not ready to embrace a widespread use of crypto-assets and blockchain technology for all the financial services that today are offered worldwide. Second, there is still a considerable number of financial services which cannot be substituted by protocols. These are the ones that still require an important human component to be performed.

As for the others, there may be a slow substitution, although initially, some substitutes may not be fully "unpolluting". Metaphorically speaking, in choosing a transportation method, we do not opt for the horse because the horse is less polluting than the car. We still choose the car, although we are doing our best to spread the adoption of electric cars.

References

Allen TFH, Hoekstra TW (1993) Toward a definition of sustainability. General Technical Report RM (USA) 247:98–107

Aratani L (2021) Electricity needed to mine bitcoin is more than used by 'entire countries'. https://www.theguardian.com/technology/2021/feb/27/bitcoin-mining-electricity-use-environmental-impact

Aspris A, Sean F, Jiri S, Leqi W (2021) Decentralized exchanges: the "wild west" of cryptocurrency trading. Int Rev Financ Anal 77:1–10. https://doi.org/10.1016/j.irfa.2021.101845

Business Standard (2019) Bitcoin causing annual CO2 emissions comparable to those of Vegas: Study. https://www.business-standard.com/article/pti-stories/bitcoin-use-causing-co2-emissions-comparable-to-las-vegas-study-119071900858_1.html

Capponi A, Jia R (2021) The adoption of blockchain-based decentralized exchanges. https://doi.org/10.2139/ssrn.3805095

Castro-Iragorri C, Ramirez J, Velez S (2021) Financial intermediation and risk in decentralized lending protocols. Cambridge Centre for Alternative Finance, Cambridge. https://doi.org/10.2139/ssrn.3893278. n.d. https://ccaf.io/cbeci/index

Chohan UW (2018) Proof-of-stake algorithmic methods: a comparative summary. https://doi.org/10.2139/ssrn.3131897

Cocco L, Pinna A, Marchesi M (2017) Banking on blockchain: costs savings thanks to the blockchain technology. Future Internet 9(3):25. https://doi.org/10.3390/fi9030025

De Hoo S, Olaerts M (2011) Sustainable development and the need for sustainable oriented corporate law and regulation. University of Oslo Faculty of Law Research Paper № 2011–29

Di Paola M (2016) "Lezione sulla Sostenibilità, lesson held at LUISS University". Youtube, March 17, 2016. https://www.youtube.com/watch?v=EJ961feV45s

Ferrarini G (2020) Corporate purpose and sustainability. *Law Working Paper* №559/2020. https://ecgi.global/sites/default/files/working_papers/documents/ferrarinifinal.pdf

Furnari SL (2018) ICO in Italia: applicabilità della disciplina sull'equity crowdfunding e suoi potenziali benefici. In: Lener R (ed) Fintech: Diritto, Tecnologia e Finanza. Minerva Bancaria, Roma

Furnari SL (2019) Validità e caratteristiche degli smart contract e possibili usi nel settore bancario finanziario. In: Corapi E, Lener R (eds) I diversi settori del fintech. Problemi e prospettive. Wolters Kluwer Italia, Alphen aan den Rijn

Furnari S (2021) Trough equity crowdfunding evolution and involution: initial coin offering and initial exchange offering. Lex Russica 1:101–117

Giungato P, Rana R, Tarabella A, Tricase C (2017) Current trends in sustainability of bitcoins and related blockchain technology. Sustainability 9(12):2214. https://doi.org/10.3390/su9122214

Guo Y, Liang C (2016) Blockchain application and outlook in the banking industry. Financ Innov 2:24. https://doi.org/10.1186/s40854-016-0034-9

Harris WL, Wonglimpiyarat J (2019) Blockchain platform and future bank competition. Foresight 21(6):625–639. https://doi.org/10.1108/FS-12-2018-0113

He P, Tang D, Wang J (2020) Staking pool centralization in proof-of-stake blockchain network. ERN: Other Game Theory & Bargaining Theory

Investopedia.com (n.d.). https://www.investopedia.com/terms/c/consensus-mechanism-cryptocurrency.asp

John K, Rivera TJ, Saleh F (2020) Economic implications of scaling blockchains: why the consensus protocol matters. https://doi.org/10.2139/ssrn.37504672020

Lener R, Furnari SL (2020) Company law during the blockchain revolution. The rise of "CorpTech". Open Rev Manag Bank Finance 6(2):33–52

Lu Z, Wan X, Yang J, Wu J, Cheng Z, Hung PCK, Huang SC (2019) Bis: a novel blockchain based bank-tax interaction system in smart city. IEEE Intl Conf on Dependable, Autonomic and Secure Computing, Intl Conf on Pervasive Intelligence and Computing, Intl Conf on Cloud and Big Data Computing, Intl Conf on Cyber Science and Technology Congress (DASC/PiCom/CBDCom/CyberSciTech), pp. 1008–1014

Lund EH, Jaccheri ML, Li J, Cico O, Bai X (2019) "Blockchain and Sustainability: A Systematic Mapping Study." *IEEE/ACM 2nd International Workshop on Emerging Trends in Software Engineering for Blockchain (WETSEB) (2019)*, pp. 16–23

Macchiavello E, Siri M (2020) Sustainable Finance and Fintech: Can Technology Contribute to Achieving Environmental Goals? A Preliminary Assessment of 'Green FinTech' (August 13, 2020). *European Banking Institute Working Paper Series 2020* 71. https://doi.org/10.2139/ssrn.3672989

McMaster A (2017) "Bitcoin is massively polluting the earth — and we should all be scared." https://www.globalcitizen.org/fr/content/bitcoin-pollution-environment-electricity-robots/?template=next

Medhi PK (2019) Role of blockchain enabled transparency in risk management and sustainability of the complex global supply chains. https://doi.org/10.2139/ssrn

Mohan V (2020) Automated market makers and decentralized exchanges: a DeFi Primer. Financ Innov 8:20. https://doi.org/10.2139/ssrn.3722714

Moore JE, Mascarenhas A, Bain J, Straus SE (2017) Developing a comprehensive definition of sustainability. Implement Sci 12:110

Peters G, Panayi E (2015) Understanding modern banking ledgers through blockchain technologies: future of transaction processing and smart contracts on the internet of money. https://doi.org/10.2139/ssrn.2692487

Popova N, Butakova NG (2019) "Research of a possibility of using blockchain technology without tokens to protect banking transactions." *IEEE Conference of Russian Young Researchers in Electrical and Electronic Engineering (EIConRus)*, pp.1764–1768

Rowlatt J (2021) "How Bitcoin's vast energy use could burst its bubble." https://www.bbc.com/news/science-environment-56215787

Saleh F (2021) Blockchain without waste: proof-of-stake. Rev Financ Stud 34:1156–1190
Stoll C, Klaaßen L, Gallersdörfer U (2019) The carbon footprint of bitcoin. Joule 3(7):1647–1661. https://doi.org/10.1016/j.joule.2019.05.012
Sulkowski AJ (2018) Blockchain, business supply chains, sustainability, and law: the future of governance, legal frameworks, and lawyers? Del J Corp Law 43:303–345
Thapa S, Ghosh P, Chhetri B (2019) "Security issues on Blockchain Network (BcN): a review". *2nd International Conference on Advances in Science & Technology (ICAST) 2019 on 8th, 9th April 2019 by K J Somaiya Institute of Engineering & Information Technology, Mumbai, India.* https://doi.org/10.2139/ssrn
Wang R, Lin Z, Luo H (2019) Blockchain, bank credit and SME financing. Quality Quantity Int J Methodology 53(3):1127–1140
Wu T, Liang X (2017) "Exploration and practice of inter-bank application based on blockchain." *12th International Conference on Computer Science and Education (ICCSE)*, pp. 219–224

Tokenization of the Creative Industries: The Intersection Between Emerging Technologies and Sustainability

Yulia Kharitonova

Abstract The chapter is devoted to analyzing the situation with the attraction of private investments in the creative industries for the Sustainable Development Goals. At the 74th session of the UN General Assembly, the year 2021 was declared the International Year of Creative Economy for Sustainable Development. The concept of the creative industry is based on the interaction of human creativity and ideas with intellectual property, knowledge, and technology. According to UNCTAD's Creative Economy Program, protection through Intellectual Property Rights (IPRs) is linked to opportunities to attract investment in the arts. It is possible to promote sustainable and inclusive economic growth, foster innovation, provide opportunities, benefits and empowerment for all and respect all human rights with the application of blockchain technology and the issuance of tokens. Blockchain technology and non-fungible tokens, or NFTs, have played an essential role in attracting funding to the creative industries. Tokenization in the arts and creative industries demonstrates a solution to the various challenges of Sustainable Development. The influx of investment has been linked to new technological opportunities and has also enabled additional funding or livelihoods for artists. NFTs have recently become a great sensation in the art market. Legal research has shown how NFTs are created and what rights investors can realize under smart contracts.

Keywords Sustainable development goals · Creative industries · Non-fungible tokens · Investment in creativity · Smart contract

1 Introduction

In 2019 at the 74th session of the UN General Assembly, the year 2021 was declared the International Year of Creative Economy for Sustainable Development.[1]

[1] United Nations (2019).

Y. Kharitonova (✉)
Lomonosov Moscow State University, Moscow, Russia

L. V. Sannikova (ed.), *Digital Technologies and Distributed Registries for Sustainable Development*, Law, Governance and Technology Series 64,
https://doi.org/10.1007/978-3-031-51067-0_4

Seventy-fourth session Second Committee Agenda item 17 Macroeconomic policy questions recognized the need to promote sustainable and inclusive economic growth, foster innovation, and provide opportunities, benefits, and empowerment for all and respect for all human rights. It also identified the continued need to support developing countries and countries in transition in diversifying production and exports, including in new areas of sustainable growth, including creative industries.

The term creative economy is not unambiguously defined. According to UNCTAD's Creative Economy Program, it is "an evolving concept that is based on the interaction of human creativity and ideas with intellectual property, knowledge and technology. In essence, it is a knowledge-based economic activity on which the "creative industries" are based".[2]

Creative industries include advertising, architecture, arts and crafts, design, fashion, film, video, photography, music, performing arts, publishing, research and development, software, computer games, electronic publishing, and television/radio are the lifeblood of the creative economy. Being creative in nature, these activities are also considered an important source of commercial and cultural value.

According to UNCTAD's Creative Economy Program, protection through Intellectual Property Rights (IPRs) is linked to opportunities to attract investment in the arts.

2 Blockchain for Sustainable Development of Creative Industries

Blockchain technology and non-fungible tokens, or NFTs, have played an important role in attracting funding to the creative industries. The influx of investment has been linked to new technological opportunities and has also enabled additional funding or livelihoods for artists. Museums have also found themselves interested in tokenization of masterpieces to cover their operating costs.

Blockchain technology, originally used in fintech and supply chain management, has steadily expanded into creative industries. The digital revolution has favored more developed countries, and it has helped create a digital divide with less developed countries. The introduction of blockchain into asset management can mitigate many of the critical problems, helping to improve governance and a more equitable distribution of economic benefits.[3]

NFT and blockchain technology are steadily penetrating the world of art and philanthropy, allowing artists and museums to monetize their work. As tokenization technology is new to society as a whole, the issues of implementing the Sustainable Development Goals are only just becoming predictable.[4]

---

[2]UNCTAD (n.d.)

[3]Gillpatrick et al. (2022), pp. 105–128.

[4]Umar (2022).

Thus, it can be argued that tokenization in the arts and creative industries demonstrates a solution to the various challenges of Sustainable Development:

Goal 9: Build resilient infrastructure, promote sustainable industrialization and foster innovation, target **9.3.** Increase the access of small-scale industrial and other enterprises, in particular in developing countries, to financial services, including affordable credit, and their integration into value chains and markets;

Goal 10 Reduce inequality within and among countries, target 10.2. By 2030, empower and promote the social, economic and political inclusion of all, irrespective of age, sex, disability, race, ethnicity, origin, religion or economic or other status;

Goal 1 End poverty in all its forms everywhere, target 1.4. By 2030, ensure that all men and women, in particular the poor and the vulnerable, have equal rights to economic resources, as well as access to basic services, ownership and control over land and other forms of property, inheritance, natural resources, appropriate new technology and financial services, including microfinance.

In addition to the above, indirectly, the creative industries are trying to influence a change in public attitudes toward the global agenda by promoting environmental protection. For example, against the backdrop of the pandemic during 2020, some artists have held digital art exhibitions inspired by climate change and COVID-19 (Goal 3 "Ensure healthy lives and promote well-being for all at all ages" and Goal 13 "Take urgent action to combat climate change and its impacts").[5]

Digital platforms with non-fungible tokens on which artists can create NFTs, showcase and sell their creations to raise awareness in the context of the 17 UN SDG goals include DigitalArt4Climate, the Enjin NFT platform and DoinGud, allowed to donate proceeds to fund SDG-focused charities around the world.

Axiom Zen, a for-profit company, created NFT Honu, a CryptoKitty that was inspired by sea turtles, in 2018. The start of the initiative focused on the challenge of mobilizing blockchain-based resources for marine conservation. Funds from the auction of the token were used for the conservation of sea turtles in the Caribbean.

Since increasing wealth and income inequality reduces future economic growth and limits social cohesion,[6] it has been under the research spotlight for quite some time. The solution for the creative industry is to use blockchain to manage creative assets. By using trusted technology in an untrusted Internet environment, more and more artists gain access to the investment and funding marketplace, allowing them to increase their income and wealth. In response to artists' offerings, investors and art collectors are also gaining easier access to works in the marketplace. Although the situation has now changed, in 2021 the statistics showed Trading in NFTs spiked 21,000% to more than $17 billion in 2021. The number of investors was also growing, with 2.3 million buyers in 2021 compared to 75,000 in 2020.[7] $10–20

[5] Martin (2020).

[6] Galor and Zeira (1993), pp. 35–52.

[7] Browne (2022).

million worth of NFTs are sold every week on the blockchain, with most NFTs selling for less than $300.[8]

In addition according to the analytics NFT charities that generate money for socially significant issues are often incredibly popular all over the world.[9]

Using NFTs to meet green post-COVID-19 recovery in accordance with the United Nations' 17 Sustainable Development Goals allowed Selva Ozelli to call Them "Charitable sustainable NFTs".[10]

3 Uniqueness for Achieving the SDGs

The tokenization of intellectual property is a "way of financing art", which is being actualized today to achieve sustainable development. NFTs have recently become a great sensation in the art market. From August, 31 to September, 7, 2021, a premium auction of the State Hermitage's NFT collection was held on the Binance NFT marketplace. The "Your token is kept in the Hermitage" limited digital collection was made up of digital copies personally signed by Mikhail Piotrovsky of the following works from the museum collection: Leonardo da Vinci's Litta Madonna, Giorgione's Judith, Vincent van Gogh's Lilac Bush, Wassily Kandinsky's Composition VI and Claude Monet's Corner of the Garden at Montgeron.[11] The State Hermitage Museum in 2021 announced an exhibition of works of art made in the format of Non-Fungible Tokens. According to the organizers, the Hermitage project is the first exhibition event in Russia to create and distribute NFT and their use in the art sphere and creative industries, which is the most topical issue on the contemporary art agenda today. It is believed that NFTs have become a new form of securing rights to a work of art using blockchain technology, allowing for precise "control of copyright and ownership [*of collectibles - Y.Kh.*]".[12]

Other museums are already taking their first steps toward NFT. Businessman Eduardo Burillo donated CryptoPunk NFT artwork #5293 to the Miami Institute of Contemporary Art. Thus, it is still the only museum in the world to have an NFT object in its collection. In Florence, Italy, the Uffizi Gallery recently sold a digital version of Michelangelo's «Madonna Doni» for €140 thousand at an NFT auction.[13] The Italians plan to do the same with other works from the famous collection.

---

[8] Chohan (2021).

[9] Theis (2021).

[10] Ozelli (2021).

[11] The State Hermitage Museum (2021).

[12] Mail.ru (2021).

[13] Violity (2021).

In May 2021, the Austrian Post began Crypto Stamp 3.0 pre-sale, followed by the introduction of blockchain-based collectibles—cryptocurrency stamps—into the regular sale.[14]

Hollywood's film industry mogul Sid Ganis, who has worked as a top executive at studios such as Sony Pictures, Lucasfilm, and Warner Bros, partnered with MOVE Network (sets out to be the world's largest NFT aggregator covering a wide spectrum of NFT products) said: "The film industry is constantly changing and innovating. I have been lucky to be a part of those changes for many years. Now movies and content are moving into the crypto marketplace via NFTs, which is another major shift into the 21st Century world of global entertainment. I am very happy to bring what I know to the process".[15]

Today, luxury brands (Prada, Gucci, Louis Vuitton, Balenciaga, and others) are also entering the NFT and metaverse environment with clothing sales.[16]

The NFT sales boom is due to advances in technology as well as technology investors in cryptocurrency. Collectors began buying and selling NFT artworks such as Beanie Babies in 1998. Now, art tokens are sold both through traditional auction houses such as Christie's and on specialized NFT trading platforms.[17]

What is this non-fungible investment token? NFT is a non-fungible token (virtual digital unit) as usual in a blockchain network. Generally speaking, such tokens can be conceptualized as a hash function representing a media object or as a digital certificate that confirms the ownership of a digital asset: an image, a gif-animation file, a video or audio clip, event tickets, a domain name. NFT technology makes it possible to buy and sell unique digital objects and track who owns them using blockchain. Creating a token allows a particular object to be encrypted for presentation on a blockchain network to simplify the circulation of intellectual property and other similar objects. The content of a token can be a one-of-a-kind object, such as an artwork,[18] an illustration in digital form, or a single copy of many objects, such as trading cards, and through blockchain, it is possible to keep track of who owns the NFT.

Thus, NFTs are unique hash sums stored in a blockchain and can confirm that a piece of digital art belongs to a specific person. Technically, the platform allows the artist to create NFTs and connect them, for example, to computer graphics objects created by the artist via links. There can be several tokens with the same links, but they will all be unique objects in terms of law and technology. Buyers usually get limited rights to demonstrate the digital artwork they represent, but in many cases,

---

[14] Block-chain24 (2021).

[15] Newsfile (n.d.).

[16] Encila (2022).

[17] Alter (2021).

[18] An example would be the tokenization of intellectual property, where a digital file associated with a unique token was converted into a digital file by artist Banksy Morons (White), which the blockchain company Injective Protocol bought from the Taglialatella Gallery in New York for $95,000 and then burned, organizing a broadcast.

they are simply buying "in addition to the nominal bragging rights that typically accompany an NFT purchase"[19] without the possibility of commercial use. The reference objects are (still) almost exclusively digital or physical objects that are not stored on a blockchain themselves. However, this may change in the future as a result of new technological developments, such as on-chain NFTs, which enable data to be stored on blockchains independently of trading platforms. In addition to their usefulness in making traditional digital and physical assets tradable, NFTs could also provide a basis for the metaverse.

Economic and technical requirements determine features of NFT as an object of property turnover. The following features stand out:

- programmability (NFTs are fully programable, with digital objects hosted in a smart contract[20]),
- standardization (NFT tokens have common reusable, inheritable standards, which guarantees the possibility of interaction and operating of tokens in the network),
- interoperability (NFTs must move across multiple digital ecosystems, dozens of wallets, marketplaces, and even virtual worlds),
- marketability (it is necessary to provide free selling of NFT on open marketplaces for any cryptocurrency and/or in exchange for tokens, with the possibility to use technological instruments of trade, such as auctions),
- instant tradability (instant NFT trading leads to increased liquidity of the works of art),
- immutability and provable scarcity (smart contract technologies allow to program the creation of a certain number of media objects that become rare on the market).[21]

Until recently, such tokens were associated with game objects (artifacts) in online games and digital collectibles. However, the dual nature of NFTs from a legal point of view is that they allow to fix the fact of possession of the result of intellectual activity and realize some rights to it through the execution of a smart contract.

The creation of NFT in crypto art can tokenize collectible individually defined art objects of the real or virtual world or tokenize intellectual property rights to objects created uniquely in digital form.[22] Tokenized digital art objects can be protected or not recognized as such under the current legislation. Still, the token is always a unique object due to the technology of its creation, which has value for commerce.

[19] Kastrenakes (2021a).

[20] The mechanics of NFT may not be limited to a simple smart contract model, but include such technologies for working with content as forging, crafting, random generation, etc.

[21] Fairfield (2021).

[22] Sannikova and Kharitonova (2020).

4 NFT in the Art Market: Is it Appropriate to Talk About Art? Legal Approaches in Russian and International Law

To release an NFT, it is necessary to connect to a digital platform and place the content (media file) itself in the token program. The media file can be a picture, a video, a 3D model, audio. The name of the token should be fixed. Various characteristics can then be added to the token, as well as locked content that can only be seen by the buyer (the owner of that token). As a rule, tokens are created as a single copy, but this is not obligatory. NFTs can be sold on platforms in an auction format.

The most popular approach to defining the legal regime of NFT in the arts and entertainment today is to point to its collectible value. There are several main categories of NFT traded on crypto-exchanges:

- art: paintings, images, gifs, photos;
- sports: cards, videos, equipment;
- games: characters, cards, costumes;
- domain names;
- metaverses;
- words and tags;
- music;
- memes.

Of course, there are NFTs of real estate and other unique things and goods. But this area is not the focus of this chapter. The media content of a non-fungible token in the arts is based on an object that belongs to copyright objects or on means of individualization. As a rule, the work is initially expressed in digital form. But sometimes, a piece of art is also taken as the basis of NFT. In such case, a digital twin of the work is created. The digital twin is a digital copy of a physical object or process, which helps optimize the business's efficiency. This practice is widespread, for example, for the creation of digital museums. In this case, the value of the token in turnover is associated with the uniqueness of the content.

The most important emphasis is that NFT does not give direct ownership of those objects that have a material component in real life. Generally, the purchase of an NFT only grants the purchaser ownership of the specific copy or version of the work that the NFT represents. A cryptographically signed receipt that you own a unique version of a work.

The laws of many countries confirm this. For example, under German civil law, according to the very predominant view, NFT as type of crypto assets does not constitute absolute rights and, due to their lack of corporeality, they are not "objects" (Sachen) within the meaning of sections 90 and 903 of the German Civil Code (Bürgerliches Gesetzbuch - BGB). Likewise, it is not possible to transfer ownership of a crypto asset through agreement and transfer of ownership, according to sect. 929 of the BGB.

General rules of Article 1227 of the Civil Code of the Russian Federation establish that intellectual rights do not depend on property rights or other real rights to a material medium (thing), in which the corresponding result of intellectual activity or means of individualisation is expressed. So the transfer of ownership of a thing does not entail the transfer or granting of intellectual rights to the result of intellectual activity or to a means of individualisation expressed in this thing.

Going back to the examples of tokenization of masterpieces of world art in manuscripts, we are convinced that the purchase of an NFT only grants the purchaser ownership of the specific copy or version of the work that the NFT represents. A cryptographically signed receipt that you own a unique version of a work. Ownership of the NFT underlying asset does not pass to the token owner unless it is fixed in the contract.

In this context, an NFT right can be defined as having the nature of a copyright. But it is still not clear for the lawyers whether the rules on copyright can be extended to digital assets. First of all, copyright law protects the underlying asset. The uniqueness of tokenized media content copyright is so essential for the marketplaces that some of them (like SuperRare and Foundation) require not to mint multiple NFTs for the same artwork.

One would assume that the NFT is a copyrighted work. At the same time, if we consider the issue of content protection through the prism of copyright, the feature of uniqueness or novelty as a criterion of protection is not reflected in Russian law. On the contrary, the mere absence of novelty, uniqueness, and (or) originality of the result of intellectual activity may not indicate that such result was not created by creative work and, therefore, is not an object of copyright (paragraph 80 of the Rulings of the Plenary Session of the Supreme Court of the Russian Federation № 10, April 23, 2019 "On the Application of Part IV of the Civil Code of the Russian Federation"). According to Russian legislation, a particular result of intellectual activity within the copyright and related rights framework is recognized as a protected object if it is created by creative work. At the same time, it must be born in mind that until proven otherwise, the results of intellectual activity are presumed to have been created by creative work.[23] In this case, the uniqueness and scarcity in the market of non-fungible tokens is associated not with the legal criterion of intellectual property protection, but with the laws of the so-called collection design (unique, one-of-a-kind objects of art).[24]

Emphasis is placed on the uniqueness of the protected object, which enables the buyer to obtain an investment income in the future through the subsequent sale of the digital asset. Artists and designers have realized that the mass production of cultural objects, whether photographs, clothes designs, or furniture, leads to mass fakes. Therefore, there is a move away from serialism to art object design in the art world: an individual object, or an object produced in small series (capsule collections), is

[23] The Rulings of the Plenary Session of the Supreme Court of the Russian Federation № 10, April 23, 2019 "On the Application of Part IV of the Civil Code of the Russian Federation".

[24] Proctor (2020).

much easier to protect from offenders. Since copying happens at an enormous speed in the digital environment,[25] creating NFT based on content allows to ensure the connection of original content with a unique hash of a token.

Thus, the underlying asset is unique and original as an object of copyright. NFT is also unique: not as a copyright object, but as a unique hash amount obtained by the blockchain algorithm.

Further, the smart token contract can include both restrictions on content replication and the prohibition of content use outside of the copyright holder's rights. They are stored on the blockchain and contain various terms and conditions (for example, the artist's royalties when the work is resold). This approach would not solve the issue of infringement of digital content that remains online outside of the token shell. But tokenization of copyright objects in this manner is just aimed at sufficient identification of individual-defined objects in the virtual space. NFT technology allows protecting the interests of copyright holders through technical characteristics of tokens, though it does not exclude unauthorized copying. The purpose of limited art tokens is usually not to impose more substantial restrictions on use or a new form of digital rights management but to create new types of selling digital assets. Customized copies of the original content are also allowed, which is also reflected in the token's smart contract. NFTs often allow users to use digital content, whether posting it on social networks, using it as an avatar, etc.

In the sphere of tokenization of art it is also customary to include game tokens and tokens of metaverses. NFTs are used in games with in-game purchases to confirm ownership of virtual items such as armor sets, weapons, buildings, and skins. Primarily, players can transfer items between their accounts, resell them, and use tokens to prove ownership. Players do not have exclusive rights to these game objects. The rights still belong to the developer company, but the NFT token confirms the indisputable right to these collectibles. So that players can transfer items between their accounts, resell them, and use tokens to prove ownership.

But we cannot say that all tokens are related to the digitalization of the turnover of protected results of intellectual activity and means of individualization in the context of Part Four of the Civil Code of the Russian Federation. Firstly, domain names are not objects of intellectual property rights in Russia. However, trade-in NFT accounts and domains are ubiquitous.[26] Secondly, one of the most promising approaches, ideally suited for NFT technology, in the event industry. Selling tickets has always been a costly process involving fraud, authenticity issues, and reliance on a monopoly of distributors. The smart ticket Real Madrid Football Club NFT enables select fans and partners to receive exclusive rewards on their digital smart ticket. The individuals can then redeem digital currency that they can use to access exclusive digital items designed specifically for the event.[27]

---

[25] Entin (2017).

[26] See, for example: Rare and exclusive NFTs for blockchain domain names available for the first time https://unstoppabledomains.com/premium-domains-nft-sale.

[27] TASS (2021).

NFT offers solutions to all of these problems and many others.[28] The We.new platform offers NFT a recording of Scottish tennis player Andy Murray's 2013 Wimbledon victory speech (a limited-edition digital file representing the meaningful moment of the speech, "because that's all NFT really is").[29]

Thus, NFT in the arts is not always tokenization of exclusively protectable intellectual rights objects. But because of tradition, such objects of circulation also belong to the art industry.

Because NFT is not a copyrighted work, it may be considered a derivative work i.e., works being a remake of other works. Someone who creates an NFT using someone else's work should ensure they have permission from copyright owner. Copyright law provides a "bundle of rights" which are exclusive to the owner of the copyright in a work. However, usually the copyrights on the derivatives have no value and are not transferable, NFTs extract value in other ways.

It should be noted that Russian law has the legal concept of digital rights as objects, the content, and conditions of exercise, which are determined according to the rules of the information system. Execution, disposal, including pledge, encumbrance of a digital right in other ways, or restriction of disposal of a digital right is possible only in an information system. Under Article 141.1 of the Civil Code of the Russian Federation, such objects as digital rights must be explicitly named in the law. Only in this case will tokens be subject to legal rules. However, direct regulation is only addressed to such digital rights, which concern the tokenization of debt and corporate governance in financial markets. Legislation adopted in Russia has not mentioned anything about the tokenization of intellectual property and creative industries. The minimal scope and depth of special regulation in this sphere force lawyers to select a suitable legal regime to protect and defend the rights to tokenization of the art market.

The Russian legal approach allows to divide NFTs into security tokens and utility tokens depending on their functions. Though NFTs commonly represent art, some NFTs have digital "utility." These NFTs give owners rights to perform certain actions on a blockchain. For example, the online horse racing game "ZED" sells such tokens. ZED is "a digital horse racing game where players can build a stable of racehorses by buying, breeding, and racing digital racehorses." Buyers purchase NFTs that represent digital "horses." Each digital horse has a breed, blood type, and other characteristics affecting racing potential.[30] The value of this NFT derives partly from its utility in addition to scarcity and collectability.

We believe that in relation to the objects of art the legal regime of NFT as an object containing copyrights is disclosed through the fixation of rights and the content of the IP licenses.

[28] Breed (2021).

[29] Howcroft (2021).

[30] ZED RUN: Digital Horse Racing. n.d. https://zed.run/terms.

5 Securing Property Rights in Blockchain

Securing both real estate property and intellectual property rights is of great importance for economic development and welfare. As art becomes digital and documents go online, new technologies are that blur physical boundaries emerging. Tokenization in this context is a technology, an innovative way to register copyrights and rights to marketing designations (trademarks, commercial designations, domain names).[31]

Haydaroğlu in his analysis of the international property rights index found a positive relationship between the degree of property protection and foreign investment in EU and OECD countries.[32] Canals and Şener[33] found that the guarantee of intellectual property rights is an important determinant of investment, especially in high-tech industries. Therefore, we consider the application of blockchain technology for the formalization and circulation of rights to creative assets via NFT to be very fruitful.

Fixing information about the NFT owner allows publishing that the asset belongs to a particular person but limits the "double spending" of the digital artifact. At the same time, the token code displays information about the creator of the NFT similar to the information about the authorship and the right holder. In the absence of legalized registration of intellectual property results in blockchain, the creation of a token is not proof of recognizing the author's non-property and other intellectual rights. In terms of law and order, only a court of law can establish authorship in case of a dispute.

As for determining whether copyright belongs to any person, everyone should bear in mind that in Russia, there is only optional registration of copyright with the Russian Authors' Society (RAO), other authorized organizations, or digital platforms like IPChain. Such organizations do not check whether the object being registered is protected by copyright. The author guarantees upon signing the statement that he "is the sole owner of the intellectual property rights to this work and that the rights of others have not been violated in the creation of the work. In the statement the author states that "the legal meaning of the fact of deposit of the result of intellectual activity and conditions of deposit are explained, including the fact that this deposit does not affect the occurrence, implementation and protection of the result of intellectual activity (copyright), but can be used by the author as evidence of the fact of the existence of unpublished result of intellectual activity at the time of its deposit in RAO".[34] In characterizing the registration of works in RAO, its functional, technical nature is noted: it helps find and identify works and their authors.[35]

[31] Sudarikov (2010).

[32] Haydaroğlu (2015), pp. 217–239.

[33] Canals and Şener (2014), pp. 17–31.

[34] Russian Authors' Society. n.d. http://rao.ru/for-rightholders/deponirovanie-neobnarodovannyh-proizvedenij/.

[35] Novoselova and Ruzakova (2017), pp. 334–349.

This registration confirms only the existence of a specific registered object and has no legal value. Such registration does not confirm the applicant's authorship or primacy in the creation of the object.

The system of distributed lager registers (including blockchain technology) allows quickly and at a low cost without intermediaries to fix the time of existence of the intellectual property and the existence of the right holder. The application of such a certification method is recognized as effective for artworks, but it is also admissible for industrial property objects.

According to the Decree of the President of the Russian Federation from 28.06.2021 N 378 "On creation of public-state organization "Russian Center for the Circulation of Rights to the Results of Creative Activities" the basis for legalization of the information and telecommunication infrastructure of the rights management on the results of creative activity, including in scientific and engineering sphere, created with a view of promotion of the circulation of such rights based on digital technologies was laid down. Thus the development of the regulatory framework is going towards legalization of the use of information technologies (including the end-to-end digital technology of distributed registers) for the management of rights to the results of intellectual activity. At the same time, the emphasis is placed on the possibility of creating technological systems of rights management.

The mentioned ways of confirming the work presented to an independent person at a particular moment (notarial or other) are aimed only at the fixation on this fact. Still, they do not save the participants of circulation from the necessity to perform many other actions to sell their objects as well. On the contrary, thanks to the creation of NFT, the procedure for the transfer of intellectual property results or rights to them is reduced to a single step—the purchase of a token. Specialized trading platforms have been created for the creation, accounting, and circulation of digital collectibles created by tokenization and secured by intangible assets—the results of intellectual activity.[36] The use of blockchain technology on some platforms also allows tracking the history of an object, changes in its NFT value, and other details.

In the absence of an effective and reliable mechanism for copyright protection, NFT responds to market demand. Thereby digital platforms are designed to protect the authorship of right holders. At the same time, the law does not establish but requires the development of conditions to verify persons registering NFT and their rights to media content. In the absence of requirements explicitly stated in the law, the only thing that one of the most famous digital platforms Rarible offers is verification of the name and social networks of the person registering an NFT in his name.

---

[36] However, there is still an acute problem of legalization of such acts in the blockchain network: Kharitonova (2019), pp. 15–21.

6 Tokenization Is a Way to Put Digital Art into Circulation and Protect Its Rights

NFT sales are conducted using a smart contract. The efficiency of trading works and other digital assets in NFT is achieved by establishing balanced smart-contract terms. The vast majority of NFT smart contracts are compared to IP licensing agreements.

But this is not always the case. For example, the co-founder and lead vocalist of Linkin Park, Mike Shinoda, sells NFTs under the following "NFT Terms" which are located on his website: "Only limited personal non-commercial use and resale rights in the NFT are granted and you have no right to license, commercially exploit, reproduce, distribute, prepare derivative works, publicly perform, or publicly display the NFT or the music or the artwork therein. All copyright and other rights are reserved and not granted".[37]

Blockchain and smart contracts add several unique properties to non-fungible assets that fundamentally change the user-developer relationship with this type of asset. Technologically, blockchain provides a coordination layer for digital assets in the form of tokens, enabling users to manage assets through smart contracts.

The creation of NFT makes it possible to simplify and accelerate the disposal of a token without concluding a formal licensing agreement or an agreement on the alienation of an exclusive right. In particular, the right of use may be limited in terms of territory or time or to certain types of use. Of particular importance in this context is the distinctions between non-exclusive or the far-reaching exclusive rights of use which give their buyer exclusivity. Transfer of a token as a digital asset with an independent value on the market is not always associated with the transfer of exclusive rights to the result of intellectual activity. It is not without reason that we are often talking about the creation of a collection. At the same time, achieving the goals of sustainable development, philanthropy, and the green economy can be established under the terms of NFT licenses.

The terms of purchase of digital art objects can vary with the agreement of the buyer and the seller, depending on the situation. Within the capabilities of the digital platform on which the realization of NFT takes place, the parties themselves can agree on the terms of the purchase and sale of NFT and use the platform as a tool to ensure the reliability of settlements and the transfer of tokens. The transaction can be structured according to the model of acquisition of real objects of art, or the parties can dispose of the exclusive rights to the corresponding object. But even after the sale, the NFT-object remains on the Internet: anyone can download it, print it out and hang it on the wall as long as the NFT holder owns the original. Also it could, for example, be agreed in the code of a tamper-proof smart contract that an author will receive a proportion of the proceeds if the buyer resells for a certain price. There is

[37] Kerrang (2021).

also the option of defining that a share is held in every further transaction. In these cases, no further human act of transposition is required.

Tokenization of unique objects solves the problem of obtaining protection for not very original objects. In addition, "originality" is a very forgiving standard for granting and protection to the results of intellectual activity, the researchers note, but it does not include ideas. And many works of contemporary art consist of nothing more than common elements and ideas that copyright cannot protect.[38] Thus, conceptual art usually includes an abstract idea that is not protected by copyright. Thereby tokenization allows to overcome the limitations of intellectual property rights and to fix the ownership of both legally protected and unprotected objects of intellectual activity. For example, Dapper Labs, operator of the popular NBA Top Shot platform, has promulgated a template NFT License that NFT sellers can adopt to outline what rights are being licensed to the NFT buyer. The NFT License distinguishes the actual NFT token from the "art" associated with the NFT (the underlying image, music, sound, or combination thereof). The license clarifies that the buyer of the NFT receives (1) a personal license to use and display the art associated with the NFT, as well as (2) a commercial license to make merchandise that displays that art associated with the NFT, a license subject to a $100,000 gross revenue per year limit.[39]

Creating tokens with non-fungible soft assets makes it possible to determine the fate of exclusive rights to the tokenized object. Attracting investment in art is possible by directly granting rights to the use of the intellectual property by several token holders and to the results of intellectual activity that exist as unique assets that can be the reason for attracting investment by sharing the rights of the holders. We are talking about the division of the right between several right holders—token holders, which seems to be the most promising for the tokenization of the sphere in question. Intellectual property assets, such as licenses for works, trademarks, patents, as well as royalties from music and media rights, generally have low liquidity and currently have almost no secondary market for trading. In this sense, the concept of tokenization of intellectual property implies a "split right": to the result of intellectual activity as such or a split right to that result of intellectual activity (expressed in digital form) rather than an intellectual right.

From a legal point of view, the purchase of a token as an investment object does not allow for proof of ownership of an intellectual product. The winner of an auction on an NFT trading platform receives tiny: basically "a digital file plus some fuzzy image rights."[40] That said, the smart contract does not always reflect the transfer of rights to use the NFT object, such as the various ways to physically display the work, whether it is on TV in the token holder's home or on the Internet. A token is just a representation of something, whether it's a picture or a song. The copyright holder will dispose of the copyright in the traditional form by making a contract, albeit in

[38] Frye (2021).

[39] Medium (2019).

[40] Kastrenakes (2021b).

digital form. The possibility of license agreements based on smart contracts is widely discussed by lawyers but is not yet in demand in practice.

7 Conclusions

Features of tokenization of intellectual property in this case are associated with its inherently intangible nature. The intangible nature of intellectual property as an object of civil turnover combines very well with the speed and extraterritoriality of turnover of digital assets based on blockchain.

We are talking not only about the sale of the collectible object, but also about the possibility of "sharing" the right between several right holders - holders of tokens, which seems the most promising for expanding the investment attractiveness of the sphere in question. The economic effect of tokenization of intellectual rights lies in the possibility to provide financing for creative industries in various ways.

Asset-backed tokens can represent "exclusive and irreplaceable assets, such as collectibles".[41] And are therefore distinguishable from one another by design. Accordingly, each token would be unique, which would entail digital scarcity because blockchain network participants know how many tokens are in circulation and their distinctive features (unique characteristics).

Essentially, the platform allows the artist to create NFTs and link them, for example, to computer graphics objects created by the artist. There can be several tokens with the same links, but all of them will be unique objects in terms of law. The platform allows the parties to sell and buy NFTs in the future and acts as a technological organizer of such an exchange. The parties can negotiate the terms of NFT purchase and sale themselves and use the platform as a tool to ensure the reliability of settlement and token transfer. The transaction can be structured on the model of acquisition of real objects of art, or the parties can dispose of exclusive rights to the corresponding object.

Attracting funding to creative industries allows solving social problems, which contributes to achieving sustainable development goals.

References

Alter R (2021) "Screen Time at the 'World's First Physical NFT Gallery,' Whatever That Means". *Vulture*. April 2, 2021. https://www.vulture.com/2021/04/superchief-first-physical-nft-art-gallery-whatever-that-means.html

Block-chain24 (2021) "Australia Post launches new cryptocurrency postage stamps". June 25, 2021. https://www.block-chain24.com/news/novosti-nft/pochta-avstrii-zapuskaet-novye-

[41] Hobbs (2018).

kriptovalyutnye-pochtovye-marki?utm_source=yxnews&utm_medium=desktop&utm_referrer=https%3A%2F%2Fyandex.ru%2Fnews%2Fsearch%3Ftext%3D

Breed C (2021) "NFTs are ready to disrupt the ticketing world!" Medium. May 24, 2021. https://medium.com/centrality/nfts-are-ready-to-disrupt-the-ticketing-world-5cf13c637912

Browne R (2022) "Trading in NFTs spiked 21,000% to more than $17 billion in 2021, report says". CNBC. March 10, 2022. https://www.cnbc.com/2022/03/10/trading-in-nfts-spiked-21000percent-to-top-17-billion-in-2021-report.html?&qsearchterm=NFT

Canals C, Şener F (2014) Offshoring and intellectual property rights reform. J Dev Econ 108:17–31

Chohan UW (2021) Non-fungible tokens: blockchains, scarcity, and value. *Critical blockchain research initiative* (CBRI) Working Papers

Encila J (2022) Prada joins balenciaga and gucci Web3 foray with own NFT launch. https://bitcoinist.com/prada-launches-nft-collection/

Entin VL (2017) Copyright in virtual reality (new opportunities and challenges of the digital era). Statute, Moscow

Fairfield J (2021) Tokenized: the law of non-fungible tokens and unique digital property. Indiana Law J, Forthcoming. https://ssrn.com/abstract=3821102

Frye BL (2021) *NFTs & the Death of Art* (April 19, 2021). 10.2139/ssrn.3829399

Galor O, Zeira J (1993) Income distribution and macroeconomics. Rev Econ Stud 60(1):35–52. https://academic.oup.com/restud/article-abstract/60/1/35/1576085?login=false

Gillpatrick T, Boğa S, Aldanmaz O (2022) How can blockchain contribute to developing country economies? A literature review on application areas. Economics 10(1):105–128. https://sciendo.com/article/10.2478/eoik-2022-0009

Haydaroğlu C (2015) The relationship between property rights and economic growth: an analysis of OECD and EU countries. DANUBE 6(4):217–239

Hobbs D (2018) Crypto collectibles: The non-fungible token craze. *TTTstudios.* May 22, 2018. https://ttt.studio/blog/crypto-collectibles-the-non-fungible-token-craze/

Howcroft E (2021) "Game, set and . . . token? Murray cashes in on 2013 Wimbledon win". *Reuters.* June 25, 2021. https://www.reuters.com/world/uk/game-set-token-murray-cashes-2013-wimbledon-win-2021-06-24/

Kastrenakes J (2021a) "Beeple and the founder of Pitchfork are launching an NFT site where you can buy moments in time". *Theverge.* June 24, 2021. https://www.theverge.com/2021/6/24/22546817/wenew-beeple-nft-pitchfork-buy-moments

Kastrenakes J (2021b) "Beeple sold an NFT for $69 million". *Theverge.* March 11, 2021. https://www.theverge.com/2021/3/11/22325054/beeple-christies-nft-sale-cost-everydays-69-million

Kerrang (2021) "Mike Shinoda has released the music from his NFT mixtape". December 6, 2021. https://www.kerrang.com/mike-shinoda-has-released-the-music-from-his-45-second-nft-mixtape

Kharitonova YS (2019) Legal significance of fixation of intellectual right by means of distributed registers technology. Law and Economics 1:15–21

Mail.ru (2021) "Money from nothing. Why the world is obsessed with NFT tokens". March 27, 2021. https://hi-tech.mail.ru/news/53455-pochemu-ves-mir-pomeshalsya-na-nft-tokenah/

Martin J (2020) "Review: Cointelegraph-Inspired Climate Change Art Exhibition". Cointelegraph. May 30, 2020. https://cointelegraph.com/news/review-cointelegraph-inspired-climate-change-art-exhibition

Medium (2019) "NFT License 2.0: Why a NFT can do what no other creative IP can do". https://medium.com/dapperlabs/nft-license-2-0-why-a-nft-can-do-what-mickey-mouse-never-could-27673d5f29aa

Newsfile (n.d.) Hollywood's Sid Ganis Joins The MOVE Network Advisory Board. https://www.newsfilecorp.com/release/85763/Hollywoods-Sid-Ganis-Joins-The-MOVE-Network-Advisory-Board

Novoselova LA, Ruzakova OA (2017) Value and functions of copyright registration in The Russian Federation and abroad. Vestnik of Perm University. Legal Sciences 3, pp. 334–349

Ozelli S (2021) "Charitable sustainable NFTs for the United Nations' 17 SDGs". Cointelegraph. September 25, 2021. https://cointelegraph.com/news/charitable-sustainable-nfts-for-the-united-nations-17-sdgs

Proctor RA (2020) "Is Collectible Design a New Form of Contemporary Art? Nomad St. Moritz/3rd edition". Flash – art. March 5, 2020. https://flash%2D%2D-art.com/2020/03/is-collectible-design-a-new-form-of-contemporary-art-nomad-st-moritz-3rd-edition/

Sannikova LV, Kharitonova YS (2020) Digital assets: a legal analysis. Moscow

Sudarikov SA (2010) The right of intellectual property. Moscow

TASS (2021) "Real Madrid and LAVA deliver first-ever Smart Ticket NFT experiences on Flow". November 3, 2021. https://tass.com/press-releases/1357469

The State Hermitage Museum (2021) "Results of the "Your token is kept in the Hermitage" auction". September 8, 2021. https://www.hermitagemuseum.org/wps/portal/hermitage/news/news-item/!ut/p/z1/pZJNb8IwDIb_yjhwrOykSVuOEeuKYAwx1kFzqboQlU5tCiyi-_kLTNMO-0BjOUSOYuuxX78gYQXSFIeqLGzVmqJ270wG-UyIgPhDHCcTDFAkSGdiSv0kDuARJMjalJBpcwy3qlpDxviaBYXPPcUD4rEIuVdopT2uFGU8xCeCeMxWxm7tBrKN3jeVLUp9pU0fVWusNraPRncvHzdFSt7jnAzCnBJYnutNum_84Qh09fJcSuYQ4ScCUxajiBZ3JLxlBCcUlodKd5Cadt84tRZ_VGD0hZDcJNduiMF8GLEZxZR_R1gOp_k8je9d7WXKn4GG_xxrfJL118U4V1XPu50UzgHHXb9aWF1kgW3TRH7jrdDLxoeuexg1-TT2eV2KXu8NMZfNyg!!/dz/d5/L2dBISEvZ0FBIS9nQSEh/?lng=en

Theis M (2021) "Charities wade into NFT craze with mixed financial results". December 29, 2021. https://apnews.com/article/technology-business-beeple-philanthropy-7741a0266f9c0ff2bb8464d5481af1b3

Umar A (2022) Metaverse for UN SDGs-An Exploratory Study. In Proceedings of the 7th Multistakeholder Forum on Science, Technology and Innovation for the SDGs, May 5–6, 2022. https://sdgs.un.org/sites/default/files/2022-05/2.1.4-27-Umar-Metaverse4SDG.pdf

UNCTAD (n.d.) Creative Economy Programme. https://unctad.org/topic/trade-analysis/creative-economy-programme

United Nations (2019) Seventy-fourth session Second Committee Agenda item 17 Macroeconomic policy questions. https://undocs.org/A/C.2/74/L.16/Rev.1

Violity (2021) "Michelangelo Buonarroti's painting turned into NFT tokens was sold for 170 thousand dollars". March 19, 2021. https://violity.com/en/new/5162-michelangelo-buonarroti-s-painting-turned-into-nft-tokens-was-sold-for-170-thousand-dollars

Part III
The Green Digital Transformation: Legal Issues

The Importance of a Global Legal Framework and Digital Technologies in Combating Climate Change

Cocou Marius Mensah and Tjaša Ivanc

Abstract This article discusses the impact of United Nations (UN) conferences on environmental protection, specifically the 2015 Paris Climate Agreement (PCA) and its effects on sustainable technologies. Various regulations to fight against global warming have pushed for reform in several sectors, including economy, energy, agriculture, and education. This article analyses the changes in those sectors and introduces the European Green Deal as a climate action plan and the latest IPCC report on the state of the environment. The paper deduces that the new technologies implemented to reduce global pollution have positive results on the environment and the economy of the pioneer states, which are already positioned for the leadership of new market shares.

Keywords EU green deal · Paris agreement · Sustainable development · Climate change · Climate action · Kyoto protocol · Doha agreement · Conference of parties · New technologies

1 Introduction

This article aims to properly situate the impact of the Paris 2015 Climate Agreement[1] (PCA) on climate in the international legal arsenal and its role in promoting sustainable technologies. No summit of the last decade could have been more critical than that of the 2015 Paris conference, given the stakes, the necessity, and the importance of the agreement. The event mobilised heads of states and governments to sign the deal to prompt the worldwide fight against global pollution and mostly global warming. This agreement was geared towards achieving success in emissions

[1]UN Climate Change Conference (COP21) in Paris. Accessed February 10, 2023. https://www.un.org/en/climatechange/paris-agreement.

C. M. Mensah (✉) · T. Ivanc
University of Maribor, Maribor, Slovenia
e-mail: cocou.mensah@um.si; tjasa.ivanc@um.si

L. V. Sannikova (ed.), *Digital Technologies and Distributed Registries for Sustainable Development*, Law, Governance and Technology Series 64,
https://doi.org/10.1007/978-3-031-51067-0_5

reduction and sustainable development, exploring the horizon of renewable energies and improving digitalisation. The PCA was not novel by any means as the 1997 Kyoto Protocol brought together 37 industrialised countries to reduce greenhouse gas emissions to limit global warming. However, when juxtaposed to its predecessor, it should be noted that the Paris agreement proved to be more efficient due to the drastic differences between the responsibility and involvement of States under the Kyoto Protocol.

The ineffectiveness of the Kyoto protocol was underscored by the withdrawal of countries such as the United States, Canada or Japan. Conversely, with the PCA being aimed at limiting greenhouse gas emissions and imposing a pollution limit for all states globally, it may be deemed the ultimate effort to fight against global warming. To achieve the objectives of keeping the global average temperature below 2 °C, a concrete legal basis and regional or continental initiatives are necessary to decarbonise the environment and limit the pollution caused by human activities. A total revolution of all layers of society is thus essential. We have to rethink the mobility of people and goods, tourism, daily routines, energy production, the fossils fuel industry and even our hobbies. In line with the UN's sustainable development goals (17),[2] the PCA can only be fully achieved by developing and implementing green technologies.

However, in 2017 the United States announced its withdrawal from the PCA, as it did with Kyoto.[3] Nevertheless, it is clear that for economic and strategic reasons, it is almost impossible for the United States to abandon the promising market of electric cars and sustainable infrastructure because the market share of renewable energies and new technologies is soaring.[4] Complying with the PCA is undoubtedly beneficial for the environment but also represents an economic opportunity. Indeed, digitalisation and the sustainability of every aspect of our society are essential to achieve sustainable development goals. PCA changes the entire social paradigm in the sense that its compliance requires a green economy, a green automotive industry, a resurgence of digitalisation in all aspects of our daily life, and finally, digital currencies to ensure the online payment of goods and services. In short, with the protection of the environment and the fight against global warming, we are entering the era of responsible consumption, mobility without emissions and the option of non-polluting technologies and energy resources. It is a new reality with new markets and technologies taking hold. Missing this opportunity to be a leader in the era of digitalisation and green technologies will be fatal from an environmental point of view and dramatic from an economic perspective.

It is not fair to say that the PCA is the only driving force behind the development of new technologies in recent years. However, it would also be remiss not to

[2] The UN Sustainable Development Goals (2023).

[3] Press statement on the U.S. Withdrawal from the Paris Agreement. Accessed February 10, 2023. https://2017-2021.state.gov/on-the-u-s-withdrawal-from-the-paris-agreement/index.html.

[4] Center for Climate and Energy Solution (n.d.).

recognise the crucial role that this environmental agreement continues to play in our daily lives due to its political, socio-economical, and cultural influence.

This article will analyse how the determination to save our planet by reducing greenhouse gas emissions has participated in developing new technologies and new market shares.

2 Climate Conferences And Their Legal-Political Impact

It is undeniable to mention the role played by the UN climate conferences in protecting the environment. The series of talks known as the "Conference of the Parties (COP)" actively influenced the gradual development of the global plan to define methods, mechanisms and means to fight against global warming.[5] It must be said, that the United Nations succeeded in holding conferences focusing on climate, a mark of its effectiveness in leading the institutions and countries towards an agreement that arranges stability and sustainability. In this sense, one cannot speak of conferences on the climate without citing two meetings that had a significant impact on the regulatory mechanisms and the exact objectives to be achieved. These are COP III with the Kyoto protocol[6] and COP XXI with the Paris agreement. Paris held the UN climate talks from November 30 to December 11, 2015, and that meeting was nearly cancelled due to the terrorist attacks in Paris, which killed at least 129 people. The urgency of the climate issue prompted the French government to maintain its position, thus limiting the event to negotiations between parties and without side events, such as manifestations.[7]

The new mechanism of climate politics introduced in Paris had clear objectives. Namely, to succeed in finding an international consensus involving all states, including the less developed and the developed, and to share the climate responsibility. The aim is to avoid repeating the Kyoto protocol errors, which have been criticised for setting unattainable targets and timetables for emission reductions.[8] In order to encourage countries to reduce their emissions and take the necessary actions to save the blue planet and ensure sustainable development, the Paris agreement had a double mission in the sense that it encompasses avoiding global warming due to human activities and ensuring sustainability for the future generations. Maintaining the global temperature increase during this century well below 2 °C above pre-industrial levels and continuing efforts to limit the temperature rise even further to 1.5 °C had become a goal to achieve absolutely with the collegial participation of

[5] United Nations Climate Change (n.d.-a).

[6] The Conference of the Parties (COP 3) (1997).

[7] Howard (2015).

[8] Böhringer (2003).

all the countries since globalisation has also played a role in consumption growth and international activities.[9]

The scientific basis for this decision lies in the reports of the IPCC[10] which, since its creation in 1988, has produced reports which have identified human activity as one of the factors of global warming. As we know, industrialised countries spend a lot of energy to support productivity and run their industries, factories, and economies. This causes them to rely heavily on traditional energy sources such as coal and hydrocarbons. Fossil fuel energy sources have terrible consequences on the environment and actively participate in greenhouse gas production, identified as one of the elements responsible for global warming. Therefore, it is urgent to invest in other sources of energy (renewable energies, for example) to fight global warming.[11] The task appears difficult because it could affect the world economy and our daily routine, so several countries procrastinate.

Long before Paris 2015, and long before the 1992 8 Millennium Development Goals of the United Nations in Rio de Janeiro, during the Earth Summit, the United Nations Framework Convention on Climate Change was open for signature in an early effort to tackle climate change.[12] The convention, which entered into force in 1994, aimed to "stabilise the concentrations of greenhouse gases in the atmosphere at a level which prevents any dangerous anthropogenic disturbance of the climate system". It has been widely seen as a cornerstone in the fight against CO_2 pollution, and its outcome has been well documented in the scientific world with several publications on the topic, for example, Sitarz (1993),[13] Palmer (1992),[14] and Freestone (1994).[15] Since then, the countries that are parties to this convention have met periodically to discuss measures to implement climate change. Other environmental agreements such as the Kyoto Protocol have entered the legal arsenal of the fight against CO_2 with unfavourable economic mechanisms for polluting countries, intending to push them to reduce their pollution or dependence on polluting energies.[16] Countries that participated in the first round of Kyoto, such as Japan, New Zealand, and Russia, did not set new targets in the second commitment period. Canada (withdrew in 2012) and the United States (has not ratified it). Despite the lack of interest of some countries in the Kyoto Protocol, the threshold for its Amendment called the Doha Amendment 2012 was nonetheless reached on October 28, 2020, when 147 Parties had deposited their instrument of acceptance.[17] The Amendment entered into force on December 31, 2020.

---

[9] Croucher (2018).

[10] IPCC (n.d.-a).

[11] Ogunrinde et al. (2018), pp. 103–111.

[12] United Nations Treaty Collection (1992).

[13] Sitarz (1993).

[14] Palmer (1992), pp. 1005–1028.

[15] Freestone (1994), pp. 193–218.

[16] Barrett (1998), pp. 20–39.

[17] United Nations Climate Change (n.d.-b).

However, the difficulty with the Kyoto mechanism is its lack of consensus. Discussions between parties began and will continue in 2021 but are complex, involving politics, finances and the lack of consensus among parties. This is why the 2015 Paris Agreement followed a different path, preferring a broader consensus and, above all, a strategy based on sustainable development. Indeed, the effects of global warming, climatic anomalies, hot winters, the drying up of rivers in places, floods, and untimely storms in other corners of the globe, have ended up sensitising public opinion to the imminent action to save future generations. In addition, a great challenge at the international level is the battle for the leadership of the green technologies market which is profiled because the demand for green products and green energies have increased worldwide. With this in mind, the bilateral agreement between the United States and China was concluded on November 12, 2014, to stimulate green technologies and reduce their respective greenhouse gas emissions by 2025 for the first and 2030 for the second.[18] It should be noted that the role of these two countries is essential in the sense that they alone represent 42% of total CO_2 emissions and are therefore major contributors to pollution.[19] It is therefore unthinkable to fight against global warming without the active participation of these two countries. However, history repeated itself with the withdrawal of the United States from the Paris agreement as they did with the Kyoto protocol.[20] At first glance, with the history of American withdrawals from different conventions, we would say that the Paris conference received a blow. However, this was not the case. If the economic and environmental impact of the US withdrawal from the Kyoto Protocol[21] caused the failure of that agreement, the picture is totally different when it comes to Paris 2015.[22] Indeed, thanks to new technologies and widespread awareness, regulations to decrease energy dependence and state assistance to support green technologies are flourishing.[23] Curbing climate change and fighting against growing environmental pressures is the mission that the European Union, for example, and other organisations have set for themselves. To meet the environmental challenges is to meet the demand on limiting air pollution, one of the leading causes of mortality in Europe,[24] the loss of biodiversity, deforestation, waste management etc.

However, without a deadline to implement the necessary changes, no results would be achieved. A rapid transition to a green economy, green energy, green infrastructure, green transportation and a complex educational system considering climate challenges is needed. For that reason, the European Union launched the

[18]Taylor and Branigan (2014).

[19]Blokhin (2022).

[20]BBC (n.d.).

[21]Hamasaki (2002).

[22]The Climate Reality Project (2017).

[23]OECD (2019).

[24]European Environment Agency (2021).

Green Deal, an agreement between member states to make Europe the first continent without CO_2 emissions by 2050.

3 The EU Green Deal As an Outcome of the Paris Agreement

Since climate change and the crises associated with rising temperatures will change our societies in all aspects, it is necessary to propose a solution that radically changes our customs. That said, we cannot fight against pollutants without changing energy policies or financing the companies that depend primarily on the hydrocarbons business. Therefore, the pillars which make up society require a radical transformation, a change based on innovation and green technology.

The facts are indisputable. According to the communication from the Commission to the European Parliament, the European Council, the European Economic and Social Committee and the Committee of the regions on The European Green Deal, one million of the eight million species on the planet are at risk of disappearing.[25] Our means of transportation, industries, consumption habits and human activity largely contribute to deforestation, pollution, and the deterioration of aquatic life. The consequences of the above activities will have terrible impacts on the European and world economy at all levels. Already the respiratory diseases due to pollution, the costs of care, and the impact on the standard of living of the populations are clearly indexed. The socio-cultural, political and economic implications of global warming no longer leave rapid action in doubt. To implement change, the European Council on December 13, 2019, and the European Parliament on January 15 2020, have supported Europe's newest and the world's most ambitious project to tackle pollution and access green transition. The new European Green Deal policy set aims to make Europe the first-ever climate-neutral continent by 2050. Though the initiative will not be implemented in Poland[26] as the country has stated that it will not be ready by the deadline (2050) and that "it will be reaching climate neutrality at its own pace", the project has been accepted by other member states and many countries outside the EU are implementing policies similar to the Green Deal.

The Green Deal is a European realisation of the promises of COP XXI and a solid foundation to sustainable development in all of the member states committed to reducing their greenhouse gas emissions and keeping global temperature increase below 2 ° C by 2100. It is a step forward for the European Union (EU) and its citizens. The Green Deal will foster in Europe a green economy, green energy, green technology, sustainable tourism, sustainable industry and transport, and promote sustainable investments. By 2050, Europe is scheduled to be climate neutral, and its

[25] EUR-Lex (2019).

[26] Poland strongly depends on the fossil fuels industry and 80% of its electricity come from coal. See: Skoczkowski et al. (2018), pp. 1303–1349.

economy and energy adapted to sustain green development. The EU green agreement will be binding because it will be incorporated into the first European climate law.[27] The aim is to textualise in this law the objectives set in the European Green Agreement and make the players in the fight against global warming more responsible. All EU countries must reach the bar zero greenhouse gas emissions by 2050. In the future, all actions, investments, project financing and European policies will consolidate the European climate law. Since 2050 is the target year for achieving these goals, the systematic actions to reduce public and even private investments in fossil fuels or polluting technology across the EU must be urgent.

Undoubtedly, the Green Deal and the implementation of the European climate law indicate the new European growth strategy for the next 30 years. The primary focus of this strategy is the good of the citizens, the transformation of the European territory into a green space conducive to commercial activities, and perfectly combining green energy with the circular economy to promote sustainable projects.

The Green Deal strategy also aims to bring together the population of the European Union, estimated at 446 million inhabitants by Eurostat. It is a visionary project which immediately traces the common path to be followed for the next 30 years. The program ensures European growth and solidarity, which will affect all areas of the EU's management.

The implementation of the ambitious program aims to reduce pollution and tackle global warming by developing a European made green economy, green technology, sustainable industry, and transport. The policy is compatible with the UN's 17 sustainable development goals to transform the world. Innovative technologies have a significant role in both initiatives: the Green Deal and the UN Sustainable development goals since most objectives require innovation and evolutive technologies. For example, in order to achieve goal number 3 (good health and well-being), goal number 4 (quality education), goal number 6 (clean water and sanitation), goal number 7 (affordable and clean energy), goal number 9 (industry, innovation and infrastructure) goal number 11 (sustainable cities and communities) for instance, up to date technologies, political will, public awareness and climate action are necessary. Such initiatives cannot be based on the same economic system that promotes fossil fuels.

Moreover, a mechanism of assistance and transition training is important for a smooth and effective implementation. For that reason, the financial pillar of the plan is based on the "Investment Plan" and the "Just Transition Mechanism" to enable greener infrastructures, to improve the quality of air, water and soil, produce healthier food and sustainable agriculture. Investments in the program will require a massive €1 trillion in sustainable investments over the following decades and at least €100 billion in assets over the period 2021–2027 to support workers and

[27] The European climate law entered into force on July 29, 2021. https://climate.ec.europa.eu/eu-action/european-green-deal/european-climate-law_en#:~:text=of%20the%20economy-,Formal%20adoption,force%20on%2029%20July%202021.

citizens of the regions most impacted by the transition through the "Just Transition Mechanism", which will target a fair and just green transition.[28]

Overall, the green deal will mobilise nearly 25% of the EU budget for climate financing and invest in environmental objectives through several EU programs in the coming decade, reaching a staggering €503 billion long-term budget by 2027. The initiative will be financed by the EU Member States, public and private investments mobilised by InvestEU and the ETS funds (Modernisation and Innovation Funds). The Green Deal aims to prioritise sustainability and the well-being of citizens (art.3 of the Treaty on European Union) thus, aligning itself with the 2030 UN Sustainable goals as the centre of European policymaking strategies and milestones.

The policy will impact all the industries, business regulations, investments, and policymaking in the EU without any doubt. European funding will only be favourable to green projects and everything related to sustainable development such as smart cities, sustainable education, sustainable tourism, sustainable energy and the like.

Though the fossil fuel industry and other polluting activities are still well present in the structure of the European economy; it is worth noting that the Green Deal will have an overall short- and long-term positive impact on the EU, as it will:

- Modernise the European economy
- Reduce emissions to promote concrete action and fight climate-related problems
- Make the EU an undisputed sustainable champion and a world leader on on climate issues
- Make the EU the epicentre of green investments and sustainable strategies
- Initiate, develop, direct and promote new trends in the circular economy
- Leverage green investments
- Export European expertise in the fight against climate change.
- Promote European Green Quality standards.
- Promote Green tourism
- Create green technologies
- Stimulate green employment and job offers overall
- Boost member state's ability to face climate change
- Prepare the EU against any climatic eventuality and boost citizens' confidence. With its policy implementation, the Green Deal will address the twin challenges of sustainability and digital transformation. A circular economy action plan will help modernise the EU's economy and benefit from the circular economy opportunities domestically and globally. Digital technologies: artificial intelligence, 5G, cloud and edge computing, and the internet are critical enablers for attaining the Green Deal's sustainability goals in many different sectors. The sector of renewable energy is vital to the achievement of the deal. It basically addresses all aspects of societal life and activities, from transport, which accounts for a quarter

[28] European Commission (n.d.-a).

of the EU's greenhouse gas emissions, to construction with smart infrastructure to ensure sustainability.

There will be new business opportunities for all the sector operators regarding the food value chain due to the high demand for sustainable food. The Farm to Fork Strategy will strengthen efforts to tackle climate change, protect the environment, and preserve biodiversity in the fishing and agriculture sector. Awareness-raising and education have an essential role to play to achieve targeted results.

In the previous paragraphs, we have highlighted the EU's plan to achieve climate neutrality. However, as already mentioned above, results cannot be achieved without technological innovation at all levels. The role to be played by the latest development in technologies in the energy, agriculture or automotive sectors, for example, is crucial.

4 Impact of the Paris Agreement On Digitalisation and Education

We firmly believe that without goal number 4 of Goals of sustainable development, it would be difficult, if not impossible, to achieve all the other goals.[29] Clearly, education is the key to sustainable development, the knowledge that will make it possible to achieve all the objectives set. To meet this challenge, the Johannesburg Plan of Implementation (JPOI) has proposed principles to ensure universal primary education already in 2015 and the elimination of gender disparities in primary and secondary education (target planned to be achieved in 2005) and at all levels of education (target planned to be achieved in 2015) according to the Dakar framework for action.

Ensuring inclusive and equitable quality education and promoting lifelong learning opportunities for all is essential, but unfortunately the task is not yet fully accomplished. Especially with the development of digitalization and artificial intelligence, tools are needed, and the need to equip classrooms with modern equipment to ensure the quality of education, shows that there is still a lot to be done, especially in developing countries. Several authors have rightfully mentioned the place of education when it comes to sustainable development.[30] In addition to representing moral imperatives, preserving the needs of future generations and ensuring social justice, the principles of sustainable development impact the economy, trade and industry, politics and international relations,[31] which are all elements taught at the academic level. Education is one of the flagship sectors that should undergo changes linked to innovative technologies to combat climate change and improve living

[29] SDG#4-Quality education. https://sdgs.un.org/topics/education.

[30] Bebbington and Unerman (2018), pp. 2–24.

[31] Holden et al. (2017).

standards through awareness-raising. For example, in legal sciences, the educational tools and didactics are called to change to face a modernised teaching approach that is more digital.

In the White Paper on the future of Europe,[32] the Commission suggested that the destiny of Europe's future depends on its capacity to change and adapt in the next decade. Indeed, digitalisation will affect all sectors of life, starting with households, services, and professions. It is, therefore, strategic to take advantage of the digital transformation in Europe to encourage awareness-raising of every citizen. At the same time, businesses and universities are crucial to help prepare the practical steps needed to achieve the goals. The European digital plan considers not only education but also all sectors of activity and development. Europe wants to make digital technology a tool[33]to improve the lives of its citizens. It is a technology for the benefit of the citizen and not against them. Digitalisation must be able to, by enhancing productivity and automation, facilitate work and make the economy competitive and fair for all. Therefore, it is a strategic plan that intends to position Europe as a digital hub and provide it with the tools to benefit from the new trends.

The EU's single market fit for the digital age is a powerful economic and political strategy that will contribute to the European economy and create hundreds of thousands of new jobs. Such a holistic approach is welcomed as it deals with all the inadequacies identified in the formation. In 2010, the EU planned vocational and training policies to achieve high employability using digital tools. The Europe 2020 strategy,[34] a guideline for smart, sustainable, and inclusive growth, has been a success and has borne fruit since its launch in 2010. It has raised awareness of the need for didactic training in a constantly changing European job market. Since proper training increases the opportunity to be hired, emphasis was put on a different educational path, including vocational education and training policies to enable high employability. In this regard, for decades to come, closer collaboration with employers and better feedback from graduates would help make the study programs more relevant and best guide the choice of job seekers.

In this regard, improving the quality and relevance of higher education through high-quality and up-to-date programs implemented by universities and law schools is a step forward to meet the demands of digitalisation and the job market. Besides, the advantages already offered by the present educational system should be taken, which includes mobility, interregional and international cooperation. In this sense, international summer workshops, conferences, and meetings on digitalisation in the legal space and other themes will be useful to guide both students and teachers. Also, these exchange programs will strengthen mutual trust between higher education institutions and alleviate any concerns about mutual recognition of student skills. The study programs and implementation guidelines developed within the curricula must meet international and national requirements established by EU legislation,

[32] European Commission (n.d.-b).

[33] European Commission (n.d.-c).

[34] Eurostat (n.d.).

which will allow the subjects developed to be implemented in a uniform manner in all Member States. (Identical content, ECTS, etc.) Such measures will aim to facilitate mutual recognition of skills and encourage students to fulfil part of their study abroad obligations. It is not to exclude the training of law students in faculties of technology within the framework of exchanges, for example. All this will aim to harmonise practices and ensure that the legal profession will have hi-tech assets to better identify and respond to statutory regulations of technological innovations. Including law students in scientific research and development programs to merge law, technology and business from the ground up by uniting research centres, universities, businesses, and high-tech companies is an idea that will inevitably allow the proper adaptability of the modern legal profession.

Moreover, digital competence is among the recommendations on "Key competencies for Lifelong Learning". It includes emerging issues, such as robotics and artificial intelligence, media literacy, and critical thinking, especially concerning social media usage. The role of the higher educational institutes will be to contribute to the overall development of digital skills of the next generations of entrepreneurs, researchers and innovators. This will fill the existing gaps between legal training and digital training in digital tools. According to various data, there are significant disparities between the EU Member States regarding digital transformation performance.[35] Most legal firms and most practitioners lack the skills to harness digital technologies. The EU Digital Education Action Plan prioritised improving education systems through better data analysis and foresight. The EU Commission supports the initiation of strategic vision on key trends in digital transformation for the future of education systems. According to the Digital Transformation Scoreboard 2017, legal counselling is valuable resource to support the industry in digital transformation challenges.

To achieve success and ensure the digital transformation process is integrated into the legal formation, it is essential to:

- tackle the shortfalls that legal education is facing in transmitting technological skills to students since there is a need to incorporate the skills and the knowledge necessary for a competent digital legal professional. Law schools need to enable their students to develop awareness and understanding of current trends and new technologies.
- Encourage the cooperation of academics from law and information technology to ensure that digital—legal knowledge can be embedded in the curriculum. Modern legal education should offer opportunities to learn about the technology that impacts legal service delivery and to understand (theoretical, technical, practical knowledge) the necessary science behind new technologies.
- Support Erasmus+ by contributing to the cooperation between Higher Education Institutions (HEI) and industry, both of whom are directly involved and linked in global technology trends.

[35] European Commission (n.d.-d).

- Refresh teaching methodologies by adapting them to existing and emerging technologies relevant to the delivery of legal services.
- Facilitate the design and implementation of new syllabi based on interdisciplinary subjects, accessed through open access to educational resources.

According to the European Commission[36]: "In the future, 9 out of 10 jobs will require digital skills. At the same time, 169 million Europeans between 16 and 74 years – 44% – do not have basic digital skills". Traditional jobs will be transformed and reinvented under the digital influence. The prerogatives of certain professions and the qualities necessary for access to full employment must then be redefined by academies and universities to meet present and future expectations. One of the main problems of global legal education is its lack of digitalisation of the learning process. It is well-known that most law schools do not actively prepare students to use digital tools in their profession. This, in return, has a negative result as most law students are unable to yield the profitable digital market, which requires new skills. It is a fact that universities and educational centres fail to predict the evolutionary curve of new technologies and their implication in societal relations. It is nearly impossible to integrate new updated activities in the traditional as the trainers do not fully master the effect or the technological tools.

For this reason, it will be necessary to ensure that the training of jurists is not exclusively attributed to law professors but also open to technicians and actors of technical change in society. The proposed approach should be multidisciplinary and multidimensional to adjust legal training to the legal needs of a rapidly changing society. A part of the solution is to see law schools become more engaged, simulation-based and focused on active learning and climate action.

5 Future Perspectives in Enviromental Protection And New Technologies

COP 27 in Sharm el-Sheikh in November 2022 will go down in history as the summit where a crucial agreement on financial assistance to developing countries exposed to the dangers of climate change was reached. The question of how to regulate the aid to countries vulnerable to climate disasters is recurrent, but a significant step forward was made at COP 27 when world leaders decided to set up new financing mechanisms, as well as a specific fund, to help developing countries to deal with loss and damage.[37] The previous COPs have of course paved the way for this achievement, especially thanks to the IPCC report that COP 26 clearly highlighted. The 26th United Nations Conference of the Parties on

[36] Tallinn Digital Summit (2017).

[37] The Sharm El-Sheikh Climate Change Conference (2022).

Climate Change[38] (COP26), organised by the United Kingdom, in partnership with Italy, in Glasgow from October 31 to November 12, 2021, will have the mission of distilling the report of the IPCC, published on Monday, August 9 2021. The report is critical to the global situation with climate change. It also came at the right time, just a few months before the Glasgow conference, which is supposed to raise awareness on climate change and strict compliance with the 2015 Paris agreement. In Dublin, a preparation session was held: "The first Dublin climate dialogues"[39] with John Kerry, US President's Special Envoy for Climate, to support and encourage the parties before the summit. The United States re-joined the Paris agreement[40] as no country will be spared by the devastation linked to global warming due to emissions. In the logic of its predecessors, the Glasgow conference provides for a declaration on how to materialise the commitments of Paris 2015 by implementing the energy policies and actions to be adopted to reduce greenhouse gas emissions and safeguard the planet. However, needless to say, that the fight against climate change is a fight against time, the consequences of which exceeds our generation. The goal of the Paris agreement was to maintain the warming at 1.5 ° C desperately. In that case, this threshold may not be reached by 2050 according to the IPCC latest record, no matter how much humanity tries to reduce carbon pollution.[41] This unfortunate scenario will lead to heatwaves, rising ocean levels, melting glaciers, floods, the disappearance of territories, the advance of the sea, erosion, etc. Currently, at 1.1 °C of warming, the effects are already dire yet, the IPPC provided for optimistic and catastrophic scenarios, depending on the reaction of States and respect for climate agreements.

A solution to fasten the green transition is investing in green energy sources. New classes of investors are taking the challenge, for example, automotive manufacturers investing in electric cars to ensure their market share. Electric vehicles will contribute to reducing emissions and constitute a capital market, given the 0 emission policies in the EU. Green technology-producing companies will benefit from the green transition challenge to reduce emissions and control the green market share. The employment sector will be impacted by the prospect of the long-term replacement of polluting and dangerous industries such as coal mines or hydrocarbons by sectors powered by green energy.

A new working class, a healthy society and a resizing of the economic sector are offered, at the same time in the fight against global warming. Therefore, it is correct to link the new green economy to the environment because we are called upon to take drastic actions to rethink the economic fabric, activities, investments, and lifestyles. Another example of such measures is the decision of the US president

[38] *UN Climate Change Conference UK* 2022 (n.d.).

[39] Dublin Climate Dialogues. https://dublinclimatedialogues.com/.

[40] Press statement on the United States officially Rejoining the Paris Agreement https://www.state.gov/the-united-states-officially-rejoins-the-paris-agreement/#:~:text=On%20January%2020%2C%20on%20his,becomes%20a%20Party%20again%20today.

[41] IPCC (n.d.-b).

to boost the auto industry in his action plan to tighten environmental regulations and encourage a green vehicle fleet. Like the EU, the United States plans to establish itself in the green automotive market by 2030 to produce more environmentally friendly cars. Half of the vehicles sold in the United States by 2030, or less than 10 years, must be emission-free, electric, plug-in hybrids or hydrogen.[42]

Investing in the green economy and supporting the development of a new range of cars that are respectful of the environment is a challenge that will positively impact the economy of the countries involved. It is therefore not surprising to see that China, the EU and the United States are starting to invest heavily in renewable energies to ensure a better climate transition and at the same time ensure their economic hegemony. The share of the electric car market in 2020 accounted for only 2% of new car sales in the United States—against 10% in Europe, overtaking China as the biggest market.[43] Therefore, the fight against global warming inevitably involves a battle for the domination of the green automobile industry and all the better because the planet will benefit from it.

6 Conclusion

The 2015 Paris climate conference was, let's say, a tool for global awareness of climate change and its effects on our daily lives. Deforestation, the rise in sea level linked to the melting of glaciers, non-renewable energies, polluting infrastructure and means of transport have ended up being the major causes of global warming. Even the sceptics realise that mitigation of global warming will impact our society's different aspects, such as jobs, education, and our way of life.

Governments have an essential role in the process by implementing climate-related decisions. Awareness-raising, education and a more digital lifestyle could help achieve emissions reduction targets and limit global warming, thus complying with the UN sustainable development goals. Adaptation to earth-friendly energy sources and sustainability is mandatory by using new technologies and the switch to digitalisation. The process is already underway, and all sectors of activity are involved. The digitalisation of education and its effect on some jobs is an example mentioned in this article as the start of the transition process. Regulations to ensure sustainable development and treaties between nations to reduce emissions demonstrate the importance of the mission. Protecting the environment, ensuring the sustainability of resources for the next generations, and the planet's climate stability has become global priorities. It is necessary to highlight that not a single government or organisation can claim to solve the equation of climate change or global warming in autarky. Despite all the plans and programs to reduce emissions to 1.5 ° C or at least below 2 °C, new technologies have proven essential for implementing

[42] Capital (2021).

[43] IEA (n.d.).

environmental policies. Without innovation and the application of new technologies in the paradigm of ecological problems, it is impossible to find an adequate solution. The United Nations has already made the achievement of the 17 goals of sustainable development a priority. The European Union reinforces these priorities through its Green Deal. The commitment to sustainable development and the limitation of emissions is nowadays more relevant than ever. Given the circumstances and the effects of global warming, a radical change and a shift to a more sustainable way of life is required.

References

Barrett S (1998) Political economy of the Kyoto protocol. Oxf Rev Econ Policy 14(4):20–39. https://doi.org/10.1093/oxrep/14.4.20

BBC (n.d.). https://www.bbc.com/news/science-environment-54797743

Bebbington J, Unerman J (2018) Achieving the United Nations sustainable development goals: an enabling role for accounting research. Account Audit Account J 31(1):2–24. https://doi.org/10.1108/AAAJ-05-2017-2929

Blokhin A (2022) "The 5 Countries That Produce the Most Carbon Dioxide (CO2)". *Investopedia.* January 31, 2022. https://www.investopedia.com/articles/investing/092915/5-countries-produce-most-carbon-dioxide-co2.asp. Accessed 1 Aug 2021

Böhringer C (2003) The Kyoto protocol: a review and perspectives. Oxf Rev Econ Policy 19(3): 451–466. https://doi.org/10.1093/oxrep/19.3.451

Capital (2021) Le président américain veut que les Etats-Unis accélèrent enfin la transition vers un parc automobile plus vert. August 5, 2021 https://www.capital.fr/auto/voiture-electrique-joe-biden-vise-50-de-vehicules-sans-emissions-dici-2030-1411440

Center for Climate and Energy Solution (n.d.). https://www.c2es.org/content/renewable-energy/

Communication from the Commision to the European Parliament, the European Council, the Council, the European Economic and Social Committee and the Committee of the Regions. The European Green Deal (COM/2019/640 final). https://eur-lex.europa.eu/legal-content/EN/TXT/?uri=COM%3A2019%3A640%3AFIN

Croucher S (2018) Globalisation and belonging: the politics of identity in a changing world. Rowman & Littlefield, Lanham

Dublin Climate Dialogues. https://dublinclimatedialogues.com/

EUR-Lex (2019) Communication from the Commision to the European Parliament, the European Council, the Council, the European Economic and Social Committee and the Committee of the Regions. The European Green Deal (COM/2019/640 final) https://eur-lex.europa.eu/legal-content/EN/TXT/?uri=COM%3A2019%3A640%3AFIN

European Commission (n.d.-a) The Just Transition Mechanism: making sure no one is left behind. https://ec.europa.eu/info/strategy/priorities-2019-2024/european-green-deal/finance-and-green-deal/just-transition-mechanism_en

European Commission (n.d.-b) White paper on the future of the Europe. Reflections and scenarios for the EU27 by 2025. https://ec.europa.eu/transparency/regdoc/rep/1/2017/EN/COM-2017-2025-F1-EN-MAIN-PART-1.PDF

European Commission (n.d.-c) Shaping Europe's digital future. https://ec.europa.eu/digital-single-market/en/content/european-digital-strategy

European Commission (n.d.-d) The digital economy and society index (DESI). https://digital-strategy.ec.europa.eu/en/policies/desi

European Environment Agency (2021) Air pollution: how it affects our health. December 21, 2021. https://www.eea.europa.eu/themes/air/health-impacts-of-air-pollution

Eurostat (n.d.). https://ec.europa.eu/eurostat/web/europe-2020-indicators)
Freestone D (1994) The road from Rio: international environmental law after the earth summit. J Environ Law 6(2):193–218
Hamasaki H (2002) The economic and environmental impact of the US withdrawal from the kyoto protocol. https://www.gtap.agecon.purdue.edu/resources/res_display.asp?RecordID=957
Holden E, Linnerud K, Banister D, Schwanitz VJ, Wierling A (2017) The imperatives of sustainable development: needs, justice, limits, 1st edn. Routledge, Milton Park. https://doi.org/10.4324/9780203022177
Howard E (2015) "COP 21: Paris climate change conference 2015: Organisers of cancelled Paris climate march urge global show of support". The Guardian. May 19, 2015. https://www.theguardian.com/environment/2015/nov/19/organisers-of-cancelled-paris-climate-march-urge-global-show-of-support
IEA (n.d.) Trends and developments in electric vehicle markets. https://www.iea.org/reports/global-ev-outlook-2021/trends-and-developments-in-electric-vehicle-markets
IPCC (n.d.-a) Climate change widespread, rapid, and intensifying – IPCC. https://www.ipcc.ch/2021/08/09/ar6-wg1-20210809-pr/
IPCC (n.d.-b) Climate Change 2021: The Physical Science Basis https://www.ipcc.ch/report/ar6/wg1/
OECD (2019) Innovation and business/market opportunities associated with energy transitions and a cleaner global environment, February 2019. https://www.oecd.org/g20/summits/osaka/OECD-G20-Paper-Innovation-and-Green-Transition.pdf
Ogunrinde O, Shittu E, Dhanda KK (2018) Investing in renewable energy: reconciling regional policy with renewable energy growth. IEEE Eng Manag Rev 46(4):103–111. https://doi.org/10.1109/EMR.2018.2880445
Palmer G (1992) Earth summit: what went wrong at Rio. Washington Univ Law Q 70(4):1005–1028
Press statement on the U.S. Withdrawal from the Paris Agreement. Accessed February 10, 2023. https://2017-2021.state.gov/on-the-u-s-withdrawal-from-the-paris-agreement/index.html
Press statement on the United States officially Rejoining the Paris Agreement. https://www.state.gov/the-united-states-officially-rejoins-the-paris-agreement/#:~:text=On%20January%2020%2C%20on%20his,becomes%20a%20Party%20again%20today
Sitarz D (1993) Agenda 21: the earth summit strategy to save our planet. EarthPress, UNKNO
Skoczkowski T, Bielecki S, Węglarz A, Włodarczak M, Gutowski P (2018) Impact assessment of climate policy on Poland's power sector. Mitig Adapt Strateg Glob Change 23:1303–1349. https://doi.org/10.1007/s11027-018-9786-z
Tallinn Digital Summit (2017) Digital Skills in Europe. https://ec.europa.eu/info/publications/tallinn-digital-summit-factsheets_en
Taylor L, Branigan T (2014) US and China strike deal on carbon cuts in push for global climate change pact. *The Guardian*. November 12, 2014. https://www.theguardian.com/environment/2014/nov/12/china-and-us-make-carbon-pledge
The Climate Reality Project (2017) "Five reasons why the US withdrawing from the Peris Agreement isn't another Kyoto Protocol mome" June 7, 2017. https://www.climaterealityproject.org/blog/five-reasons-withdrawing-paris-agreement-not-another-kyoto-protocol
The Conference of the Parties (COP 3) (December 1997). https://unfccc.int/event/cop-3
The Sharm El-Sheikh Climate Change Conference (November 2022). https://unfccc.int/cop27/auv
The UN Sustainable Development Goals (10 February 2023). https://sdgs.un.org/goals
UN Climate Change Conference (COP21) in Paris. Accessed February 10, 2023. https://www.un.org/en/climatechange/paris-agreement
UN climate change conference UK 2022 (n.d.). https://ukcop26.org/
United Nations Climate Change (n.d.-a) Conference of the Parties (COP) https://unfccc.int/process/bodies/supreme-bodies/conference-of-the-parties-cop

United Nations Climate Change (n.d.-b) The Doha Amendment. https://unfccc.int/process/the-kyoto-protocol/the-doha-amendment

United Nations Treaty Collection (1992) United Nations Framework Convention on Climate Change. https://treaties.un.org/pages/ViewDetailsIII.aspx?src=TREATY&mtdsg_no=XXVII-7&chapter=27&Temp=mtdsg3&clang=_en)

Green Finance in the EU and Russia: Legal Frameworks and Opportunities of Digitalisation

Larisa V. Sannikova

Abstract The climate agenda has recently gained new importance as global climate change threatens all of humanity. The regulation of green finance instruments needs to be improved to attract more money to fight global warming. To ensure investor confidence in green instruments, a common standard for them must be created. The European Union and Russia are forming a legal and regulatory framework to create national standards for green bonds. A comparative legal study of the EU and Russian draft laws on green finance has demonstrated similar approaches in the creation of regulatory rules. This article describes the prospects for green finance in the context of digitalisation. Based on a study of best practices (green asset wallet, digital green bonds, etc.), it is concluded that digital solutions for sustainable finance are currently still underdeveloped. However, in the future, their use will significantly increase investor confidence in green instruments and reduce costs, including through the digitalisation of the verification process.

Keywords Green finance · Taxonomy for sustainable activities · Verification of sustainable financial instruments · Digital technologies · Distributed ledger technologies

1 Introduction

Global warming caused by carbon dioxide emissions requires a major restructuring of the economies of developed and developing countries to transition to low-carbon economies. Under the impact of global climate change, in the 2015 Paris Agreement, the international community recognised the importance of "Making finance flows consistent with a pathway towards low greenhouse gas emissions and climate-resilient development" (Art. 2). Inadequate funding for combating climate change

L. V. Sannikova (✉)
State Academic University for the Humanities, Moscow, Russia

L. V. Sannikova (ed.), *Digital Technologies and Distributed Registries for Sustainable Development*, Law, Governance and Technology Series 64,
https://doi.org/10.1007/978-3-031-51067-0_6

has been highlighted by researchers.[1] Various estimates of the total annual investment required range from $580 billion to $6 trillion.[2] The leading role in financing the transition to a greener and more sustainable economy is played by governments, but this goal cannot be achieved without the efforts of private investors. Private investment in green projects is an important stimulus for the transition to a low-carbon economy. In attracting them, the regulation of green financial instruments needs to be improved to increase the confidence of responsible investors in them.

2 Challenges of Standardising Green Financial Instruments

The evolution of the green bond market clearly demonstrates the paradigm change. Until 2012, multilateral development banks have been the only issuers of green bonds. The first green bond—the Climate Awareness Bond was issued in 2007 by the European Investment Bank. In 2008, the World Bank issued its first green bond. The issuance of corporate green bonds only started in 2013. Although the green bond market is growing and has reached an estimated $2.247 trillion by beginning of 2023 according to the Climate Bonds Initiative (CBI),[3] the contribution from private investors is not significant. The European Commission reports that green bonds accounted for just 4% of total corporate bond issuance in 2020.[4]

The lack of uniform standards for green financial instruments significantly hinders the development of sustainable finance. Issuers may doubt whether their projects meet the requirements of green finance and use other financial instruments for them, which are less attractive to responsible investors. In turn, a lack of information can influence the choice of financial instruments by responsible investors.[5] Thus, the uncertainty caused by the lack of uniform standards undermines trust in the green finance market. It also promotes unfair practices, such as greenwashing.[6] Greenwashing is defined as "an unsubstantiated claim to deceive consumers into believing that a company's products are environmentally friendly".[7] Therefore, the focus is on developing international and national standards for green finance that can increase the confidence of market participants, particularly responsible investors, in green financial instruments.

---

[1] Tuhkanen (2020).

[2] Nicol et al. (2018); Volker (2021), pp. 323–330.

[3] Climate Bonds Initiative. Accessed February 15, 2023. https://www.climatebonds.net/.

[4] Theodoropoulou Sotiria (2021). https://www.etui.org/news/european-green-bond-standard-good-initiative-hanging-balance.

[5] Flammer (2021), pp. 499–516.

[6] Grene (2015).

[7] Kenton (2021).

Globally, the most widely accepted standards are the Green Bond Principles (GBP) of the International Capital Markets Association (ICMA). The GBP represent "voluntary process guidelines that recommend transparency and disclosure and promote integrity in the development of the Green Bond market by clarifying the approach for issuance of a Green Bond".[8]

Although GBP has gained recognition and becomes a guideline for most sustainable finance actors globally, regulators in many countries are making efforts to establish a national legal framework for green finance. EU and Russian authorities are working towards establishing a national legal framework for sustainable finance.

3 EU Legislative Initiatives on Sustainable Finance

Following up on the 2015 Paris Agreement, the European Commission presented on 11 December 2019 the European Green Deal, which is "a roadmap with actions to boost the efficient use of resources by moving to a clean, circular economy and stop climate change, revert biodiversity loss and cut pollution".[9] Its fulfilment would make the EU the first climate-neutral continent by 2050. The European Commission considers the financial sector as key to achieving this goal. The use of financial incentives will reorient the flow of investment towards more sustainable technologies and responsible business practices, thus contributing to a low-carbon economy.

The European Commission's action plan on financing sustainable growth includes the adoption of a package of regulations in the field of sustainable financing. In order to implement it, on 6 July 2021, the European Commission proposed a package of measures to improve capital flows to finance the transition to a sustainable economy. The package includes:

- Strategy for financing the transition to a sustainable economy;
- Proposal for a standard for European green bonds;
- Delegated act supplementing Article 8 of the Taxonomy Regulation.

The Sustainable Finance Strategy[10] identifies four main areas for action by the financial system to support the transition to a sustainable economy: transition finance, inclusiveness, resilience and contribution of the financial system, and global ambition. Funding the transition to sustainability means extending the EU Taxonomy to recognise efforts to move the real economy towards achieving climate and environmental goals; including additional sustainable activities in the EU Taxonomy; extending the framework of sustainable finance standards and labels. The inclusiveness of a sustainable financial system suggests increasing access to sustainable financing opportunities for individuals and small and medium-sized enterprises;

[8] Green Bond Principles. ISMA (n.d.).

[9] European Commission (2019).

[10] European Commission (2021).

leveraging the opportunities digital technologies offer for sustainable finance; increasing insurance coverage; supporting credible social investments; green budgeting and risk-sharing mechanisms. The sustainability of the financial sector and its contribution to sustainable development includes measures to support the financial sector in achieving the objectives of the European Green Deal, improving sustainability and combating greenwashing. Global ambitions entail promoting an international consensus on sustainable financing, including support for third countries in transition.

Through the Delegated Acts, a common system of classification of economic activities, called the EU Taxonomy, is created by establishing a list of environmentally sustainable activities. The EU Taxonomy divides projects into two categories according to the energy used: "sustainable" projects based on hydrogen, solar and bioenergy, etc., and those that cause "significant harm" to the environment using, for example, coal and lignite. The taxonomy covers sectors that account for almost 80% of Europe's direct greenhouse gas emissions. The Delegated Act supplementing Article 8 of the Taxonomy Regulation defines the content, methodology and presentation of information to be disclosed by financial and non-financial institutions regarding the proportion of environmentally sustainable economic activity in their business, investment or lending activities.

In the EU, there is a strong debate on the classification of nuclear energy and natural gas as "sustainable". The environment or energy ministers from five European countries (Austria, Denmark, Germany, Luxembourg and Spain) opposed the inclusion of nuclear energy in the EU green taxonomy.[11] Other EU countries, such as France, Poland, Slovakia and Hungary, intend to develop the sector and support its recognition as "green". Gas power plants were excluded from the first Delegated Act, but after pressure from pro-gas governments, the European Commission has announced that the issue will be addressed in the second Delegated Act. The European Commission has been forced to consider the position of Eastern European countries on this issue. Commissioner for Financial Services Mairead McGuinness has explained: "There are many countries with coal at the moment that need to make that big jump, and gas plays a role".[12]

On 1 January 2022, the European Commission announced plans to label some natural gas and nuclear energy projects as "green" investments. The Commission emphasised "a role for natural gas and nuclear as a means to facilitate the transition towards a predominantly renewable-based future".[13] The decision has drawn criticism both from environmentalists and investors.[14] However, the European Commission's position is based on scientific research. The Joint Research Centre (JRC) prepared a report for the European Commission on "Technical evaluation of nuclear energy in relation to the 'no significant harm' criterion of Regulation (EC) 2020/852

[11] Simon (2021).
[12] Euronews (2021).
[13] European Commission (2022).
[14] Amaro (2022).

(the ‘Taxonomy Regulation’)”.[15] Energy scientists emphasise the importance of natural gas and nuclear energy in achieving the SDGs, especially in an energy transition context.[16] The European Commission adopted the Complementary Climate Delegated Act on 9 March 2022. Under strict conditions, specific nuclear and gas energy activities in the list of economic activities are covered by the EU taxonomy.

The debate about recognising energy sources as green shows the complexity and ambiguity of the issue. On the one hand, classifying too narrow a range of energy sources as “green” can lead to significant problems for national economies. The energy sector is essential for economic growth and therefore for the well-being of the population. In countries where energy sources with high carbon dioxide emissions play a significant role, a drastic phase-out could lead to a drop in living standards. On the other hand, the recognition, as green energy, of sources for which there is no public consensus could undermine public confidence, including that of responsible investors, in sustainable finance.

The Proposal for a standard for European green bonds provides a so-called gold standard for how companies and public authorities can use green bonds to raise funds on capital markets to finance green projects that comply with EU taxonomies. It is voluntary and available to any private and public issuer, not only within the EU but also outside the EU. The European green bond standard (EUGBS) will allow issuers to demonstrate their financing of green projects and investors to compare the financial instruments they are offered with a credible standard. This approach reduces the risks associated with greenwashing.

The key requirements of the EUGBS are as follows:

> Taxonomy-alignment: The funds raised by the bond should be allocated fully to projects that are aligned with the EU taxonomy;
>
> Transparency: Full transparency on how the bond proceeds are allocated through detailed reporting requirements;
>
> External review: All European green bonds must be checked by an external reviewer to ensure compliance with the Regulation and taxonomy alignment of the funded projects;
>
> Supervision by the European Securities Markets Authority (ESMA) of reviewers.[17]

The two main purposes of the European Green Bonds Regulation are

- ensuring the application of uniform requirements for the use of the label “European Green Bonds” or EuGB;
- establishing a simple registration system and supervisory framework for external reviewers.

The voluntary nature of the European Green Bond Standard has caused controversial reactions when it has been discussed. Concerns have been raised about achieving the

---

[15] Abousahl et al. (2021).

[16] Safari et al. (2019), pp. 1075–1094; Aksyutin et al. (2020), pp. 463–472; Brook et al. (2014), pp. 8–16; Prince et al. (2021), pp. 47957–47972.

[17] European Commission (2021).

first stated objective. On the one hand, issuers are entitled to use the EuGB label only if they fully comply with the requirements of the European Green Bond Standard. At the same time, compliance with these requirements, experts admit, may lead to additional costs and make these financial instruments less attractive to issuers. On the other hand, issuers can label their bonds as "green" if they comply with more lenient guidelines, in particular GBP. The effectiveness of the fight against greenwashing is thereby called into question.

The fundamental difference between EuGB and GBP is the mandatory appointment by the issuer of an accredited reviewer to confirm that the green bond structure meets the EU standard and to confirm the final allocation report to be published by the issuer on the full allocation of proceeds. The external reviewer can also be third-country representatives, provided they comply with specific requirements. External pre-and post-implementation review of the issuer will increase investor confidence. According to researchers, external review is a reliable marker of bonds that actually have environmental or climate benefits, so reviewed and certified bonds are more attractive to green investors than those securities that are simply labelled "green" by the issuer itself.[18]

To the same effect, ESMA is given supervisory functions to ensure uniformity in the application of the gold standard, as well as a level playing field for external reviewers in terms of their compliance with centralised registration and supervision of their activities. ESMA is empowered to impose penalties to fulfil its supervisory tasks, such as enforcing external reviewers' disclosure requirements, stopping breaches, conducting investigations, etc.

Issuers can adapt to the new requirements before the Proposal for a standard for European green bonds comes into force. Because of the voluntary nature of EUGBS, it will be interesting to analyse the willingness of both EU and non-EU issuers to join it. It would seem that this will be an indication of the effectiveness or ineffectiveness of the measures taken. The European Commission expects that the adoption of EUGBS will provide a significant increase in green investments.

4 Development of the Legal Regulation of Green Finance in Russia

Russia significantly lags behind leading foreign countries in the development of the green finance market. Green bonds on the Russian market were first issued by a private corporation in 2018. During the last few years, the interest of large Russian companies in the new financial instrument has increased significantly.

The leading position is taken by Russian Railways, which has placed four issues of green bonds worth €500 m, CHF 500 m and RUB 100bn between 2019 and 2021. The Russian Railways Sustainability Finance Framework emphasises the

[18] Fatica and Panzica (2021), pp. 2688–2701.

compliance of the company's bond issues with GBP (ICMA) and the Climate Bonds Standard, developed by the non-profit organisation Climate Bonds Initiative (CBI).[19] External experts, both Russian and foreign, have been engaged to verify the issues.

Moscow became the first Russian region to enter the market with green bonds. The issue, worth 70 billion roubles, was placed on the Moscow Exchange in 2021. The funds from their placement are to be used to finance and refinance urban environmental projects to reduce pollutant and carbon dioxide (CO2) emissions from urban public transport. The Green Finance Committee of VEB.RF recognised Moscow bonds as a "green" financial instrument under the Russian national methodology. The compliance of Moscow's green bonds with GBP was confirmed by ICMA, which included them in its sustainability bond registry.

Despite clear progress in green finance, the Russian green bond market is not comparable to the global green bond market. For example, there are about 18 bond issues of ten issuers traded on the specialised Sustainable Development sector of the Moscow Exchange. This raises the question of what hinders the development of the Russian green bond market. This question can hardly be answered unambiguously. It seems that one of the reasons could be the insufficiently developed legal and regulatory framework for green finance.

Law-making activity in this direction has been carried out since the Presidential Decree dated 01.04.1996 № 440 "On the Concept of the Transition of the Russian Federation to Sustainable Development". The main efforts are concentrated in the field of environmental law. The development of a legal framework for sustainable financing has clearly not received sufficient attention.[20] The experts have mentioned the need to develop a regulatory and methodological framework to address green project finance and the standardisation of green bonds, including labelling issues.[21]

Activities to establish the legal regulation of green finance started with the President of the Russian Federation's instruction following the 2016 meeting of the State Council on the issue of "Environmental Development of the Russian Federation for Future Generations": "to develop, with the participation of leading business associations, and submit proposals: on the application of 'green' financial instruments by Russian development institutions and public companies".[22]

A funding system for green projects and sustainable development initiatives is currently being prepared for launch. The Russian Government adopted Decree dated 14.07.2021 №1912-r "On Approval of the Goals and Main Directions of Sustainable (Including Green) Development of the Russian Federation". It defines terms such as 'green project' and 'adaptation project', identifies priority objectives related to positive environmental impacts, and the main directions for sustainable finance. Access to special financial instruments is planned for those economic sectors listed

[19] RZD (n.d.).

[20] Gutbrod et al. (2017), pp. 44–52.

[21] Bogacheva and Smorodinov (2016), pp. 70–81.

[22] Official Internet Resources of the President of Russia (January 24, 2017).

as directions for green finance, for example, energy, construction, industry, transport. A series of regulations are to be adopted to implement this Russian government directive.

The most important of these is the national taxonomy of green projects being developed by the Ministry of Economic Development together with VEB.RF. On 21 September 2021, Decree No. 1587 of the Government of the Russian Federation "On Approval of Criteria for Sustainable (Including Green) Development Projects in the Russian Federation and Requirements for the System of Verification of Sustainable (Including Green) Development Projects in the Russian Federation" was adopted. It establishes criteria for green projects (taxonomy of green projects) and criteria for adaptation projects (taxonomy of adaptation projects).

A principal difference between the Russian taxonomy and international approaches is Russia's unconditional labelling of nuclear power as "green". A global consensus on this issue has not yet been reached. In the EU, the debate continues. In the US, the potential of nuclear power to solve the problems of climate change is also being discussed. Nuclear proponents are banking on next-generation designs, designed to be safer, cheaper, and more flexible, but critics say nuclear reactors are simply too expensive and take too long to build to be of much help with the climate crisis[23] (Parshley 2021). Russia, according to official statements, intends to push for international recognition of nuclear energy as low-carbon.[24]

Within Russia, the development of the taxonomy is also actively debated. For example, the draft of the Ministry of Economic Development classifies only hydropower plants that are damless as "green" and does not require confirmation of this status. They account for only 10% of the total number of hydropower plants. Domestic hydropower experts insist that all hydropower plants belong to renewable energy sources and should qualify as "green" for sustainable development goals[25] (Ilina 2021). Environmentalists point to the environmental damage caused by dams.

Discussions on the taxonomy, both in the EU and in Russia, clearly demonstrate the difficulty of finding a balance between environmental responsibility and national economic development. Therefore, the Russian taxonomy emphasises adaptation projects in addition to green projects. Unlike green projects, adaptation projects are not required to comply with the goals of international climate and sustainable development documents. They only have to comply with Russian environmental priorities, which are defined by the goals and main directions of sustainable (including green) development of the Russian Federation.

For the implementation of green and adaptation projects, the Bank of Russia developed regulatory rules for green and adaptation financial instruments. 1 October 2021 The Bank of Russia amended Regulation No. 706-P of 19.12.2019 "On Standards for Securities Issuance" in order to take into account the national taxonomy of green projects and national verification of sustainable financial instruments.

[23] Parshley (2021).

[24] TASS May 26 (2011).

[25] Ilina (2021).

A national verification system is provided in the Methodological Guidelines for Verification of Green and Adapted Financial Instruments and the Methodological Guidelines for Maintaining a List of Verifiers. Only legal entities included in the list of verifiers by the Methodological Centre of VEB.RF are allowed to act as verifiers. Verifiers should use the Model Methodology and Model Approaches in their activities to analyse financial instruments, including when developing their own methodologies. The VEB.RF Methodological Centre has a supervisory function in relation to verifiers. In the case of irregularities in the activities of verifiers, it has the right to issue warnings to them and, in the case of repeated irregularities, to exclude them from the list of verifiers.

Russia, like the EU, has followed the route of establishing a national sustainable finance regulation. The legal framework for green finance is in its formative stages. An analysis of the draft legal acts on sustainable financing suggests that they are generally in compliance with international approaches and market practices of leading countries.

5 Opportunities for the Digitalisation of Green Finance

Digitalisation is a global process affecting all areas of society. Research by academics and experts confirms that digitalisation can be used to improve sustainable finance.[26] Green FinTech describes phenomena emerging at the intersection of two trends: sustainable finance and financial technology. As the researchers point out, "green FinTech holds the potential to transform the existing financial system in all areas, including payments, investments, financing, advice and insurance through sustainability by, e.g., directing capital in green investments, improving data for the evaluation of green companies, etc.".[27]

Fintech solutions based on Distributed Ledger Technology (DLT), including blockchain, have the greatest potential. It is DLT using cryptographic protection that has proved to be an effective mechanism for establishing trust in an untrusted environment. The well-known advantages of this technology, such as decentralisation, transparency and immutability of transactions, and the absence of intermediaries, are all in line with the requirements of ICMA for openness, accuracy and completeness of the information disclosed by green bond issuers.

As an example, the Green Assets Wallet[28] is the blockchain-based platform for checking and reporting financial impacts. The information uploaded by the issuer and published on the platform is retained unchanged in the register, which ensures its trust by investors and the market. The Green Assets Wallet provides the option to

---

[26] Bayat-Renoux et al. (2018); Deng et al. (2019), p. 6434; Arner et al. (2020), pp. 7–35; Nassiry (2018); Sancak (2021).

[27] Puschmann et al. (2020), p. 10691.

[28] Green Assets Wallet (n.d.).

validate the data uploaded by the issuer. Validation refers to the process by which an issuer provides proof of delivery on its commitments. The issuer can download the relevant document from the validator, or request a report from a listed validator on the platform. The Green Asset Wallet is not responsible for the quality and trustworthiness of the information. The platform only provides a system for all registered users to exchange information in a transparent manner. So, in this case, blockchain does not solve the problem of eliminating intermediaries—validators.

A system involving intermediaries—external reviewers/verifiers/validators—significantly increases the cost of issuing green bonds. According to experts, issuers can expect their additional costs associated with verifier involvement to be covered by the demand for such bonds from responsible investors. However, it needs to be recognised that the system is quite costly for all actors and diverts significant financial resources away from sustainable development goals.

This problem can be solved by integrating digital technologies such as DLT, the Internet of Things (IoT), and Artificial Intelligence (AI): "Data from the real economy can be uploaded directly to the distributed ledger from sensors in a green asset and exposed to investors in real-time, demonstrating Proof of Impact. Analytical algorithms or AI can interpret messy or difficult to interpret data before it arrives in the digital wallet of the investor".[29]

The EU Strategy for Financing the Transition to a Sustainable Economy also emphasises that technological innovations, such as blockchain, IoT, Big data and AI, have a significant role to play in sustainable financing. Digital financial tools based on these digital technologies can facilitate access to sustainable finance for SMEs.

As such, a digital financial tool could be digital green bonds issued on a blockchain platform. In 2019, the Spanish banking group BBVA issued the structured green bonds to insurance group Mapfre on its own blockchain platform.[30] At present, there are many offers from technology companies and digital platforms on the market.

The BIS Innovation Hub and the Hong Kong Monetary Authority (HKMA) have developed two prototype digital platforms: the Daml prototype,[31] produced by Digital Asset in partnership with GFT, and the Liberty prototype,[32] produced by the Liberty Consortium, comprised of Standard Chartered Bank, SC Ventures, and Shareable Asset. They enable investors to invest any amount into safe government bonds, which will develop a green project. The developers promise: "Over the bond's lifetime, the investor can not only see accrued interest, but also track in

---

[29] HSBC (2019).

[30] Fariña, Pilar M. BBVA issues the first blockchain-supported structured green bond for MAPFRE. February 19, 2019. https://www.bbva.com/en/sustainability/bbva-issues-the-first-blockchain-supported-structured-green-bond-for-mapfre/.

[31] BIS Innovation Hub (2021b).

[32] BIS Innovation Hub (2021c).

real-time how much clean energy is being generated, and the consequent reduction in CO2 emissions linked to the investment.".[33]

Tokenised green bonds benefit not only investors but also issuers. According to researchers, the use of blockchain in bond issuance reduces issuer costs by more than 10 times, thereby opening up opportunities for small sustainable and green projects.[34]

However, the EU Strategy for Financing the Transition to a Sustainable Economy raises concerns about the environmental impact and increasing energy needs of data centres and distributed ledger technologies, notably for crypto-assets. The need for high energy consumption in cryptocurrency mining, especially Bitcoin, is due to the use of Proof-of-Work (PoW). However, the main problem is not the high energy consumption, but the fact that fossil fuels are used to produce it, which increases the carbon footprint. Researches show that Bitcoin has a significant impact on the environment and contributes to global warming.[35]

At present, there are two obvious ways to mitigate the negative environmental effects of activities related to crypto-assets. One is replacing the PoW consensus algorithm with another algorithm, such as the Proof-of-Stake, which reduces power consumption. The other is to switch from using fossil sources to renewable ones when mining them. According to the Cambridge Centre for Alternative Finance, as of 2020, on average 39% of PoW mining is powered by renewable energy sources, mainly hydroelectric power (Blandin et al. 2020). In any case, crypto-assets have so conquered the world that it is impossible to abandon them. It is therefore advisable to make efforts to make them more sustainable, rather than fight them.

6 Conclusion

A comparative legal study of the EU and Russian draft laws on green finance has shown similarities in regulatory approaches. The main challenges are the creation of a taxonomy for green projects, and a verification system for green financial instruments. New digital technologies, such as Distributed Ledger Technologies, including blockchain, Artificial Intelligence and the Internet of Things will play an increasingly important role in addressing these and other challenges. Digital solutions for sustainable finance based on these digital technologies will significantly increase investor confidence in green instruments and reduce costs.

[33] BIS Innovation Hub (2021a).

[34] HSBC (2019).

[35] Calvo-Pardo et al. (2021); Qin et al. (2020); Stoll et al. (2019), pp. 1647–166; Houy (2019), pp. 655–655; Foteinis (2018), pp. 169–169.

References

Abousahl S, Carbol P, Farrar B, Gerbelova H, Konings R, Lubomirova K, Martin Ramos M, Matuzas V, Nilsson K, Peerani P, Peinador Veira M, Rondinella V, Van Kalleveen A, Van Winckel S, Vegh J, Wastin F (2021, ISBN 978-92-76-40538-2, JRC125953) Technical assessment of nuclear energy with respect to the 'do no significant harm' criteria of regulation (EU) 2020/852 ('taxonomy regulation'), EUR 30777 EN. Publications Office of the European Union, Luxembourg. https://doi.org/10.2760/207251

Aksyutin OE, Ishkov AG, Romanov KV, Grachev VA (2020) The role of natural gas in Achieving Sustainable Development Goals. Int J Energy Econ Policy 10(4):463–472

Amaro S (2022) EU's plan to include gas and nuclear in 'green' ranking leaves investors confused. CNBC. January 6, 2022. https://www.cnbc.com/2022/01/06/eus-plan-to-include-gas-nuclear-in-green-ranking-leaves-investors-confused.html

Arner DW, Buckley RP, Zetzsche DA, Robin V (2020) Sustainability, FinTech and financial inclusion. Eur Bus Organ Law Rev 21(1):7–35

Bayat-Renoux F, Svensson U, Chebly J (2018) Digital technologies for mobilizing sustainable finance applications of digital technologies to sustainable finance. Sustainable Digital Finance Alliance. https://docs.wixstatic.com/ugd/3d4f2c_6767ef5b999c4e3fa42c0e05e6ea2ac3.pdf

BIS Innovation Hub (2021a) Project Genesis: Prototype digital platforms for green bond tokenisation. November 4, 2021. https://www.bis.org/publ/othp43.htm

BIS Innovation Hub (2021b) Project Genesis – Report 3. A prototype for green bond tokenisation by Digital Asset and GFT. November 2021. https://www.bis.org/publ/othp43_report3.pdf

BIS Innovation Hub (2021c) Project Genesis – Report 2. A prototype for green bond tokenisation by the Liberty Consortium. November 2021. https://www.bis.org/publ/othp43_report2.pdf

Blandin A, Pieters G, Wu Y, Eisermann T, Dek A, Taylor S, Njoki D (2020) 3rd Global cryptoasset benchmarking study. September 2020. https://www.jbs.cam.ac.uk/wp-content/uploads/2021/01/2021-ccaf-3rd-global-cryptoassetbenchmarking-study.pdf

Bogacheva O, Smorodinov O (2016) «green» bonds as an essential tool for financing «green» projects. Financ J 2:70–81

Brook BW, Alonso A, Meneley DA, Misak J, Blees T, van Erp JB (2014) Why nuclear energy is sustainable and has to be part of the energy mix. Sustain Mater Technol 1-2:8–16

Calvo-Pardo HF, Mancini T, Olmo J (2021) Machine Learning the Carbon Footprint of Bitcoin Mining. SSRN. June 1, 2021. https://ssrn.com/abstract=3886737

Climate Bonds Initiative. Accessed February 15, 2023. https://www.climatebonds.net/

Deng X, Huang Z, Cheng X (2019) FinTech and sustainable development: evidence from China based on p2p data. Sustainability 11:6434

Euronews (2021) Brussels delays decision on making natural gas and nuclear energy a green investment. April 21, 2021. https://www.euronews.com/2021/04/21/brussels-delays-decision-on-making-natural-gas-and-nuclear-energy-a-green-investment

European Commission (2019) The European Green Deal sets out how to make Europe the first climate-neutral continent by 2050, boosting the economy, improving people's health and quality of life, caring for nature, and leaving no one behind. December 11, 2019. https://ec.europa.eu/commission/presscorner/detail/e%20n/ip_19_6691

European Commission (2021) Communication from the commission to the European Parliament, the Council, the European economic and social committee and the Committee of the Regions Empty Strategy for Financing the Transition to a Sustainable Economy. https://eur-lex.europa.eu/legal-content/EN/TXT/?uri=CELEX:52021DC0390

European Commission (2022) EU Taxonomy: Commission begins expert consultations on Complementary Delegated Act covering certain nuclear and gas activities. January 1, 2022. https://ec.europa.eu/commission/presscorner/detail/en/IP_22_2

Fatica S, Panzica R (2021) Green bonds as a tool against climate change? Bus Strateg Environ 30(5):2688–2701. https://doi.org/10.1002/bse.2771

Flammer C (2021) Corporate green bonds. J Financ Econ 142(2):499–516. https://doi.org/10.1016/j.jfineco.2021.01.010

Foteinis S (2018) Bitcoin's alarming carbon footprint. Nature 554:169–169. https://doi.org/10.1038/d41586-018-01625-x

Green Assets Wallet (n.d.). https://greenassetswallet.io/

Green Bond Principles. ISMA (n.d.). https://www.icmagroup.org/sustainable-finance/the-principles-guidelines-and-handbooks/green-bond-principles-gbp/

Grene S (2015) The dark side of green bonds. *Financial Times*. June 14, 2015. https://www.ft.com/content/16bd9a48-0f76-11e5-b968-00144feabdc0

Gutbrod M, Khrapliva J, Volodin S (2017) Green bonds as a new financial instrument and prospects for their introduction in Russia. Curr Regul Curr Control 9:44–52

Houy N (2019) Rational mining limits bitcoin emissions. Nat Clim Chang 9:655–655

HSBC (2019) Blockchain. Gateway for sustainability linked bonds. HSBC, Sustainable Digital Finance Alliance. September 24, 2019. https://www.sustainablefinance.hsbc.com/mobilising-finance/blockchain-gateway-for-sustainability-linked-bonds

Ilina T (2021) Hydropower engineering in Russia will have to fight for "green" status. *Hydrotechnika* 2. https://www.hydroteh.ru/articles/add/rossijskoj_gidroenergetike_pridetsja_poborotsja_za_status_zelenoj_144/

Kenton W (2021) Greenwashing. *Investopedia*. Jan 23, 2021. https://www.investopedia.com/terms/g/greenwashing.asp

Nassiry D (2018) *The role of fintech in unlocking green finance: Policy insights for developing countries*. № 883. ADBI Working Paper

Nicol M, Shishlov I, Cochran I (2018) *Green Bonds: Improving Their Contribution to the Low-Carbon and Climate Resilient Transition*. Green Bonds Research Program Work Package 1. February 2018. https://www.i4ce.org/download/green-bonds-improving-their-contribution/

Official Internet Resources of the President of Russia (January 24, 2017). http://www.kremlin.ru/acts/assignments/orders/53775

Parshley L (2021) The controversial future of nuclear power in the U.S. *National Geographic*. May 4, 2021. https://www.nationalgeographic.com/environment/article/nuclear-plants-are-closing-in-the-us-should-we-build-more

Prince NS, Shabbir Alam M, Murshed M, Mahmood H, Ahmad P (2021) The roles of nuclear energy, renewable energy, and economic growth in the abatement of carbon dioxide emissions in the G7 countries. Environ Sci Pollut Res 28:47957–47972

Puschmann T, Hoffmann CH, Khmarskyi V (2020) How green FinTech can alleviate the impact of climate change—the case of Switzerland. Sustainability 12(24):10691

Qin S, Klaaßen L, Gallersdörfer U, Stoll C, Zhang D (2020) Bitcoin's future carbon footprint. ArXiv abs/2011.02612

RZD (n.d.). https://company.rzd.ru/ru/9972/page/103290?id=18195#main-header

Safari A, Das N, Langhelle O, Roy J, Assadi M (2019) Natural gas: a transition fuel for sustainable energy system transformation? Energy Sci Eng 7(4):1075–1094

Sancak IE (2021) Sustainable digital finance. *SSRN*. November 13, 2021. https://ssrn.com/abstract=3962698

Simon F (2021) Germany leads call to keep nuclear out of EU green finance taxonomy. EURACTIV.com. July 2, 2021. https://www.euractiv.com/section/energy-environment/news/germany-leads-call-to-keep-nuclear-out-of-eu-green-finance-taxonomy/

Stoll C, Klaaßen L, Gallersdörfer U (2019) The carbon footprint of bitcoin. Joule 3:1647–1166

TASS (26 May 2011). https://tass.ru/ekonomika/11476773

Theodoropoulou Sotiria (8 July 2021) The European green bond standard: a good initiative hanging in the balance. ETUI. https://www.etui.org/news/european-green-bond-standard-good-initiative-hanging-balance

Tuhkanen H (2020). Green bonds: a mechanism for bridging the adaptation gap? Stockholm. https://www.sei.org/wp-content/uploads/2020/02/sei-working-paper-green-bonds-tuhkanen.pdf

Volker B (2021) Green finance in Europe – strategy, regulation and instruments. Intereconomics 56(6):323–330

Part IV
Human Rights and Information Communication Technologies

Fast Internet as a Prerequisite for Sustainable and Resilient Development: Network Expansion Measures and Extension of User Rights

Bernd Holznagel

Abstract The relationship between digitalisation, sustainability and resilience is fiercely debated. However, it is largely undisputed that digital technology is fundamentally capable of supporting the implementation of the Sustainable Development Goals and must do so. During the Corona pandemic, many people were dependent on digital applications to access education, work or medical care. Thus, fast internet is a basic prerequisite for social and economic participation. However, especially in rural regions, good internet connections are not commonly available. The article presents the strategy of Germany (as an example of a large EU Member State) to provide all citizens with access to fast internet. Access to fast internet is not only a measure of consumer protection but also serves significantly to meet the Sustainable Development Goals (especially SDG 3, 4, 8, 9).

On the basis of the Directive (EU) 2018/1972 (European Electronic Communication Code), the Member States set incentives for network expansion, for example by reducing investment costs. Necessary measures are taken by the responsible regulatory authorities. One trend is to "empower" consumers vis-à-vis telecommunications providers. On the one hand, this involves transparency about the providers available on the market and the quality of their services. A number of legal instruments are available for this purpose. On the other hand, the increased importance of fast internet access for individuals has led in Germany to an enforceable "right to fast internet". In this respect, German telecommunications law goes beyond the European directive requirements and breaks new ground to achieve the SDGs.

The author would like to thank Benedikt Freese, Felizitas Heet, Christian Schepers, Florian Flamme and Jan Kalbhenn for active support and enriching discussions during the preparation of this manuscript.

B. Holznagel (✉)
University of Münster, Münster, Germany
e-mail: holznagel@uni-muenster.de

L. V. Sannikova (ed.), *Digital Technologies and Distributed Registries for Sustainable Development*, Law, Governance and Technology Series 64,
https://doi.org/10.1007/978-3-031-51067-0_7

1 Promoting the SDG Targets Through Digital Services and Networks

Among the most important guiding principles, that are currently triggering profound transformations in the economy and society worldwide, are sustainability and digitalisation. So far, the two concepts have hardly been related to each other. Digital technologies, for example, are not mentioned at all in the 17 SDGs of the 2030 Agenda. This is no coincidence. After all, they traditionally function more as amplifiers of economic processes that are predominantly based on fossil fuels and resource-consuming global freight transport. The use of electronic devices and computer networks is associated with high energy consumption and the problem of disposing of old devices.[1] Only recently, the realisation that digital technologies can and must be used to implement sustainability goals has gained ground. These connections have recently been comprehensively analysed in the main report "Our Common Digital Future"[2] by the German Federal Government's Scientific Advisory Council on Global Change. The Advisory Council is an advisory body composed of scientists that regularly prepares reports on sustainability issues. In its main report, it gives recommendations for action on how a "digitalised sustainability society" can be built.[3] It identifies important fields of application for digital technologies, including resource-saving control of energy consumption or traffic flows,[4] monitoring of ecosystems and biodiversity, or support for the development of "precision agriculture".[5]

In addition, the report identifies SDG targets that can particularly be promoted through digitalisation. The Corona pandemic showed how important e-health applications are today, especially to remote regions, as mobility was significantly limited. Digitalisation has thus helped to preserve "health and well-being" (SDG 3) in these difficult times. Another example is teleteaching to ensure equitable, quality education and promote lifelong learning opportunities for everybody (SDG 4). As face-to-face teaching was often banned at universities, teaching at universities could only be maintained with the help of a massive use of digital technologies and services. At the law faculties in Germany, it was learned in a short time how to teach the material with the help of videos, the holding of video conferences (ZOOM) and the digital provision of reading material to the students. There has also been a massive expansion of computer-assisted homeworking in workplaces and administrations during this period. Without this, the collapse in the economic performance of society would have been more pronounced. This shows that digital technologies are an

[1] Wissenschaftlicher Beirat der Bundesregierung Globale Umweltveränderungen (2019), p. 312.

[2] Wissenschaftlicher Beirat der Bundesregierung Globale Umweltveränderungen (2019).

[3] Wissenschaftlicher Beirat der Bundesregierung Globale Umweltveränderungen (2019), pp. 315 ff.

[4] Rat für Nachhaltige Entwicklung (2021), p. 38.

[5] Wissenschaftlicher Beirat der Bundesregierung Globale Umweltveränderungen (2019), pp. 372 ff.

indispensable driver for promoting sustainable economic growth and supporting the development of innovations (SDG 8, 9).

However, pandemic times also ruthlessly reveal the deficits that the internet supply has in a respective region. They have made it clear that the supply must also work in difficult life situations and thus must have a corresponding resilience. In online teaching, it could be observed on a daily basis that a considerable number of students had technical problems participating in the digital courses. Enquiries revealed that their internet connection was often not sufficiently powerful. This was especially true for students who stayed in rural areas. In the vicinity of the universities, telecommunication networks are usually well-developed, so that the teachers and the students living there were able to quickly find their way around the new digital teaching world from a technical point of view. In many cases, it also turned out that the students were provided with less bandwidth than they were promised in the supply contracts concluded with the providers. The situation was similar for many smaller companies, which had to adapt their business processes to the demands of the pandemic quickly.

To redress this and strengthen the resilience of society as a whole, the EU and its member states are pursuing a dual strategy. First, network expansion is being systematically pursued because the existing narrowband copper networks, which were laid in the pre-digital age, no longer meet today's requirements for internet provision. In addition, copper lines require considerably more energy than fibre optic lines. One of the key features of the expansion strategy is that it is based on private-sector-driven expansion. Particularly in rural regions, the expansion activity falls short of expectations. The population density there is too low to ensure that the investments are amortised. In order to create incentives for the expansion and to reduce the costs for the construction of the networks, the regulatory authority responsible for telecommunications in the Member States is taking the necessary measures (see Sect. 2).

Recently, consumers have also been empowered to assert their individual interests in this area vis-à-vis the providers (see Sect. 4). Such "empowering of the user side" also makes it clear to individuals how important fast internet access is for them.[6] If these measures succeed in catching up in broadband expansion, the performance of the entire economy and society will be improved at the same time. Germany has consistently contested this strategic approach and recently introduced an individual "entitlement to fast internet" that has received widespread media attention. This entitlement can be enforced in the courts if necessary (Sect. 6). In addition, transparency about the providers and services available on the market has been considerably improved (Sect. 9). Provisions are also made for the case that the internet speed falls short of the performance promised in the contract (Sect. 13). Before outlining essential elements of this dual strategy using the example of the regulations commonly used in Germany, the regulatory authority responsible for its implementation in Germany is to be presented. This serves to better understand the

[6] Bundesrat (2021), p. 413.

implementation mechanisms of the right to internet and the new transparency provisions. The article concludes with a conclusion.

2 Regulatory Framework for the Telecommunications Sector

The regulatory framework for electronic communications networks and services in Europe is set by the EU. The European Electronic Communications Code (EECC)[7] is decisive for this. This directive had to be transferred into national law by the Member States by 21 December 2020.[8] In national laws, such as the German Telecommunications Act (TKG), the term telecommunications is used instead of electronic communications. The lynchpin for the implementation of the regulatory framework is the national regulatory authorities. Due to European law, they are not subject to the directive rights of the national government. The member states have only limited leeway to deviate from the European requirements and to make their own regulations. Since Germany is the most economically significant country in the EU, it is cited here as a reference area.

In the area of telecommunications, the Federal Network Agency is responsible as the regulatory authority. The authority is also responsible for regulating the electricity, gas, postal and railway sectors. It has the task of implementing the objectives applicable to these sectors. Originally, the focus of the authority's activities was to ensure the promotion of competition. This was necessary because the successor to the state-operated monopoly, the German Telekom, had significant market power. After the introduction of liberalisation, competition could only be ensured if the regulatory authority guaranteed fair network access and charges. The opening of the market made it possible to significantly reduce the prices for telecommunication services and to increase their quality for the benefit of consumers. However, cases of abuse became more frequent, so that sector-specific consumer protection developed step by step.[9] Accordingly, § 2 para. 2 no. 3 TKG also stipulates as an objective of regulation the protection of user interests, in particular consumer interests, which the Federal Network Agency is to promote. It also actively strengthens the rights of consumers by providing information and supports them in solving problems and conflicts in individual cases.[10] In this way, it contributes (indirectly) to implementing the SDGs (3, 4, 8, 9) mentioned above.

On its website, the Federal Network Agency offers consumers the possibility of contacting it in an uncomplicated manner. With the Telecommunications Consumer Service, it has set up an office that deals directly with questions and complaints from

[7] Directive (EU) 2018/1972.

[8] Article 124 Directive (EU) 2018/1972.

[9] Holznagel (2018), pp. 3 (25 f.).

[10] Bundesnetzagentur (2019), p. 203.

consumers. In doing so, it also informs consumers about their rights and the obligations of service providers.[11] The Federal Network Agency's efforts to actively empower consumers are falling on fertile ground. In the telecommunications sector alone, the Federal Network Agency received 10,321 enquiries about internet and telephone in the first half of 2021.[12] Of these, 1678 enquiries concerned switching providers, 1491 enquiries concerned supply issues and 753 enquiries concerned the topics of transparency and internet speed.[13] The Federal Network Agency publishes the corresponding figures every 6 months on its website. Finally, the Federal Network Agency uses the findings from the complaints and enquiries received from consumers as a basis for applying the regulatory instruments at its legal disposal.

Support for broadband expansion has emerged as another focus of the Federal Network Agency's activities. Originally, the goal was to leave the construction of new networks exclusively to infrastructure competition. It was hoped that due to the competition between cable network providers and telecommunications network providers, networks with sufficient bandwidths would be built. This concept only worked out in the urban regions. In order to ensure nationwide provision of the most important services in the federal territory and similar living conditions, the Federal Network Agency is increasingly supporting infrastructure expansion. This is now also reflected in the TKG, where infrastructure expansion has been included as a prominent regulatory objective.[14]

Currently, the deployment status in Germany is as follows: The results of the current survey on broadband availability in Germany (as of the end of 2020)[15] show that while 95.5% of households achieve a bandwidth of more than 50 Mbit/s, access to gigabit connections (1000 Mb) remains limited with only 59.2% of households.[16] Also at European level, only 21% of the citizens have access to 5G connectivity.[17] The commitment of the authorities is driven by ambitious targets. In particular, grid expansion in the member states has priority.[18] The European Commission's Digital Compass envisages that all citizens in the EU should have access to gigabit connections and 5G connectivity by 2030. The Federal Republic of Germany has even tightened this target and wants to achieve this goal already by 2025.[19] Accordingly, regulation in the TKG has been geared towards the goal of creating gigabit-capable networks (networks with very high capacity).[20] The focus is on socio-economic

---

[11] Bundesnetzagentur (2017), p. 186.

[12] Bundesnetzagentur (2021), p. 1.

[13] Bundesnetzagentur (2021), p. 3.

[14] § 2 para. 2 no. 3 a) TKG.

[15] Bundesministerium für Verkehr und digitale Infrastruktur (2020).

[16] Bundesministerium für Verkehr und digitale Infrastruktur (2020), p. 6.

[17] European Parliament (2021), p. 1.

[18] Body of European Regulators for Electronic Communications (2021), p. 5.

[19] On possible regulatory approaches Freese (2020), p. 22.

[20] § 3 para. 2 no. 3 a) TKG.

priorities such as educational institutions or hospitals, transport hubs, businesses and farms. The programme is accompanied by extensive financial support from the federal government and the federal states. The administrations responsible there are also responsible for implementation. The Federal Network Agency, on the other hand, is endeavouring to facilitate the expansion activities of the network operators.

3 Measures to Support Grid Expansion Development of a Data Portal

First of all, the Federal Network Agency supports the network operators in realising their expansion plans by setting up a data portal. The data portal bundles information that network operators previously had to laboriously collect from individual authorities. In addition, this data is processed according to the operator's needs, so that independent processing is only necessary to a limited extent. The federal government's central information office, which is located at the Federal Ministry of Transport and Digital Infrastructure, is responsible for setting up and maintaining the data portal.[21] The ministry can, among other things, supervise the Federal Network Agency with this task.

The information to be entered into the data portal is diverse. First and foremost, there is the existing infrastructure.[22] The aim is to create a uniform image of all components that could be used for a network expansion—from cable ducts to traffic signs to underground stations.

If a region is not supplied with high-capacity telecommunications networks, the federal and state governments would like to support the expansion there in the future.[23] However, in order to know where such support is needed, it is also necessary to have an overview of future network expansion. In practice, this concerns mobile radio in particular. Accordingly, the central information office of the federal government collects every 6 months the network expansions planned by the mobile network operators for the next 12 months.[24] This survey must contain, among other things, the specific location of the planned expansion, the type of connection to existing networks and information on the expected network coverage.[25] All network expansion measures by companies and public bodies must also be listed, although the company and business secrets of the companies must be protected.[26]

[21] § 78 para. 2 TKG.

[22] § 79 TKG.

[23] Bundesrat (2021), p. 359.

[24] § 81 para. 1, 2 TKG.

[25] § 81 para. 3 TKG.

[26] § 81 para. 5 S. 2, § 80 para. 3 TKG.

In practice, information about general construction work is of particular interest to telecommunications network operators. Construction sites are of considerable benefit to network operators, as the earth has already been torn up here and the fibre optic cables required for telecommunications networks can thus be laid at low cost.

The public sector is also to be held accountable for the development of the data pool:[27] The Federal Government, the Länder and the municipalities must provide the Observatory with information on land, infrastructures and support structures owned by them.[28] All information required for the data portal must be provided.[29] In particular, location data is conceivable here, so that the properties can be used for the installation of radio masts, for example.

Finally, the data portal contains information about areas with an expansion deficit.[30] On this basis, the Observatory can ask companies and public authorities whether they are not interested in a network expansion with very high capacity in the respective area.[31] If a company then expresses its intention to carry out a corresponding expansion, the Observatory can then ask the other network operators for a corresponding intention.[32] This creates transparency for the notifying parties, who can take the information about the expansion measures of their competitors in the respective area into account in their business planning.[33] Whether this transparency measure has a positive effect depends on truthful answers from the market participants.[34] However, the TKG is limited to a "request" from the market participants and a truthful answer cannot be sanctioned with an administrative fine.[35] The effectiveness of this instrument is therefore likely to be low.[36]

4 Shared Use of Public Supply Networks

If an operator of a public supply network, such as a local gas and electricity supplier, has already laid pipes for its network, there is also an interest in making efficient use of these so-called passive infrastructures. In this way, the adverse effects on the general public, such as traffic restrictions and noise, that network expansion entails can be reduced. In practical terms, this should minimise the number of road construction sites for laying pipes in particular. A right of joint use for the passive

[27] Bundesrat (2021), p. 360.

[28] § 83 para. 1 TKG.

[29] § 83 para. 2 sentence 1 TKG.

[30] § 84 TKG.

[31] § 84 para. 2 sentence 1 TKG.

[32] § 84 Abs. 2 sentence 2 TKG.

[33] Bundesrat (2021), p. 360.

[34] Bundesrat (2021), pp. 360 f.

[35] § 84 TKG.

[36] Admittingly Bundesrat (2021), pp. 360 f.

infrastructures of public supply networks, which network operators are entitled to, is intended to remedy the situation.

In order for network operators to get a picture of existing passive network infrastructures, they are granted a right to information vis-à-vis operators of public supply networks.[37] The operator of the public supply network is in principle obliged to provide information and must provide comprehensive information on the passive network infrastructures.[38] However, a refusal of the request for information is possible with justification,[39] for example, if the provision of the information would endanger the security or integrity of the supply networks, public safety or public health.[40]

Once the passive network infrastructures have been identified, the telecommunications network operators have a legitimate interest in investigating these components. This is the only way they can determine whether the network components are suitable for shared use in network expansion. The TKG also provides for such a right to on-site investigation.[41] Here, too, there are possibilities for refusal for the operators of public supply networks.[42]

Once the information has been obtained and the on-site investigations have been carried out, the owners or operators of a telecommunications network can begin with the network expansion using the passive infrastructures. For this purpose, they are entitled to a right of joint use of passive network infrastructures of the public supply networks.[43] Co-use must be preceded by a corresponding application to the operator of public supply networks.[44] The operator of the public supply network is then obliged to make an offer on the conditions of shared use.[45] In particular, it must contain fair and reasonable conditions for shared use with regard to the price, securities to be provided and contractual penalties.[46] Here, too, the application for shared use may be rejected if one of the exhaustively listed grounds for rejection is fulfilled, such as concrete indications that the requested shared use endangers public safety or public health.[47] If the negotiations do not result in an agreement, the Federal Network Agency acts as a dispute resolution body.[48] It can also set a fair and reasonable shared use charge itself.[49]

---

[37] § 136 para. 1 TKG.

[38] § 136 para. 3 TKG.

[39] § 136 para. 4 TKG.

[40] § 136 para. 4 no. 1 TKG.

[41] § 137 TKG.

[42] § 137 para. 3 TKG.

[43] Following § 138 para. 1 TKG in the extent of § 139 TKG.

[44] § 138 para. 1 TKG.

[45] § 138 para. 2 TKG.

[46] § 138 para. 2 sentence 2 no. 1 TKG.

[47] In § 141 TKG.

[48] § 149 para. 1 TKG.

[49] § 149 para. 2 sentence 2 TKG.

5 Further Incentives for Grid Expansion

In the EU, the successor companies of the former state (monopoly) operator continue to be of particular importance for network expansion because they mostly have a position of market power in the fixed network sector. In order to promote competition, in Germany, for example, Deutsche Telekom must grant its competitors access to its networks at fixed prices. Deutsche Telekom sees this as a considerable obstacle to expansion. If access obligations existed for competitors, it would be difficult to refinance the expansion of new networks on the market. In order to create incentives for the construction of high-capacity networks, the TKG provides for the first time that the scope of access and charge obligations can be reduced in favour of binding commitments.[50] Commitments can be general cooperation commitments by the company with market power or co-investments in networks with very high capacity.

In 2020, Deutsche Telekom and EWE TEL made a commitment to cooperate. EWE TEL is a telecommunications company that operates regionally in the northwest of Germany. The jointly controlled parity joint venture (Glasfaser Nordwest) requires the two parent companies to invest 2 billion euros over the next 10 years and is expected to connect 1.5 million private households and business locations. Glasfaser Nordwest has no end-customer relations and only operates the network. Public subsidies are not used, and the parent companies are subject to a non-competition clause with regard to network expansion. It offers all TC companies non-discriminatory access on equal terms.

6 Improving User Rights Entitlement to Fast Internet Access Prerequisites and Scope of the Claim

The German Telecommunications Act of 2021 provides for the first time for a right to be provided with telecommunications services.[51] At a minimum, voice communication services as well as a fast internet access service must be available for adequate social and economic participation. In substance, this is an individualised right to contract.[52] All consumers and businesses (end users) who do not operate telecommunications networks or services are entitled to claim.[53] The opposing party to the claim is the party obligated by the Federal Network Agency to provide the services (service obligor). The claim can only be asserted by the end users at their

[50] §§ 18, 19 TKG.

[51] § 156 para. 1 TKG. For a more detailed analysis of this provision *Kafka/Eilmes-Horváth*, in: Säcker and Körber (2023), § 156.

[52] BR-Drs. 29/21, 413.

[53] Zur Definition des Endnutzers § 3 no. 13 TKG.

main place of residence or at their place of business. As a rule, the period for fulfilling the claim is 6 months.

The internet access service must allow at least email, search engines, basic online tools for education and training (e.g. a writing programme), online newspapers or online news, online purchasing of goods and services, job search tools, professional networking, online banking, use of e-government services, social media and instant messaging such as WhatsApp, as well as standard quality calls and video calling.[54] This catalogue has been expanded to include music and video streaming services (Netflix and others) as well as teleworking, based on the experience of the Corona pandemic. The latter must allow participation in both telephone and video conferences, such as Zoom or Teams.[55] Users of telework applications must be allowed secure remote access, for example, to the usual extent of cryptographically secured Virtual Private Networks (VPN) to the employer's network. This is necessary to prevent eavesdropping or manipulation of the institutions. The legislator assumes that this performance is achieved by a 30 Mbit product. The requirements for the bandwidth of the minimum provision have thus been significantly increased due to the experiences with the pandemic.[56]

When determining the technical requirements for the functioning of these minimum services, it is not only the bandwidth actually used—and not the download and upload bandwidth promised when the contract was concluded—that is decisive. Rather, the reliability (latency) with which the services can be received and sent must also be taken into account as a central parameter.[57] This is to ensure that voice is transmitted smoothly and video image files without judder. The specified values are to be usable over the entire course of the day. Further details are to be specified in a legal ordinance. This is the responsibility of the Federal Ministry of Transport and Digital Infrastructure, which, however, can transfer this task to the Federal Network Agency.[58]

7 Affordable Supply Prices

The minimum available services must be offered at an affordable price.[59] The criterion of affordability is to be specified in more detail by the Federal Network Agency. In particular, the average prices, connection costs and the orientation to the composition of the household net income, which are determined taking into account

[54] § 157 para. 3 sentence 3 TKG in connection with Annex V to Directive (EU) 2018/1972.

[55] Bundesrat (2021), p. 416.

[56] On this, see the explanatory statement of the lead Ausschuss für Wirtschaft und Energie (2021), p. 466.

[57] § 157 para. 3 sentence 2 TKG, see Ausschuss für Wirtschaft und Energie (2021), p. 466.

[58] § 157 para. 5 TKG.

[59] § 158 para. 1 TKG.

local peculiarities, serve as reference points. Within the scope of its discretionary powers, however, the Federal Network Agency is free to use criteria other than those mentioned to determine affordability.[60] The authority is also responsible for monitoring the development and level of prices for minimum supply services.[61] It is thus incumbent on the Federal Network Agency to determine an essential part of the legal entitlement to the provision of telecommunications services. The Federal Network Agency enforces consumer access to fast internet access services at affordable prices in the event of undersupply with safeguard measures. The trade-offs to be made require extensive consultation between the Federal Network Agency and consumer and business associations.[62] Further financial assistance for low-income households or disabled end-users according to the rules of social law is not excluded by this regulation.[63]

8 Identification of the Service Provider and Funding Issues

If, in the course of its supervisory activities, the Federal Network Agency comes to the conclusion that adequate and sufficient supply at affordable prices is not guaranteed, it shall declare a state of undersupply for the area concerned. If a company voluntarily assumes the prescribed minimum supply, the Federal Network Agency declares this commitment to be binding.[64] The company then acts as a service obligor against which the individual end users can direct their claims.

If no company declares its willingness to do so, the authority must oblige one or more companies to provide the minimum service in the affected area by order. However, those affected are granted financial compensation if this represents an unreasonable burden for them. At least in Germany, taxpayers' money is not used for this purpose. The costs for this compensation are refinanced through a pay-as-you-go system.[65]

All companies active in the market for the provision of voice telephony and internet access services must participate in the pay-as-you-go system. The amount of the share that a company must pay is based on the share of the annual turnover of the company in question in the total market turnover. Small, medium and micro enterprises are exempt from the levy.[66] In addition, messenger services, such as WhatsApp, which belongs to Facebook, can also be used to finance the minimum

[60] Bundesrat (2021), pp. 417 f.

[61] § 158 para. 1 TKG.

[62] Begründung des federführenden Ausschuss für Wirtschaft und Energie (2021), p. 467.

[63] Bundesrat (2021), p. 417.

[64] § 161 para. 1 TKG.

[65] § 163 TKG.

[66] § 163 para. 8 TKG.

coverage.[67] To do so, they must have a significant coverage and user base. The Federal Network Agency uses the number of monthly users in Germany as a reference point for the calculation.[68] The purpose of including these services is to spread the financing of the universal service over as many shoulders as possible. This would also benefit the end users, who would thereby experience a lower burden.[69]

9 Improvements in the Transparency of the Offers Information About Available Providers and Services Information About Network Connections at the Place of Residence

Before the pandemic, few consumers will have been aware of what their Internet access actually provides. The increased use of network infrastructures for teleworking and telelearning made it clear, especially to families with several children, how important it is today to have an internet connection designed for this purpose. For them, it is now vital to find a service designed for their needs. This is not a trivial task, as the telecommunications market is known for the confusion of providers and offers. Consequently, an information tool should be available on the market that enables consumers to find out the availability of network connections at their place of residence and thus to make a choice between several available supplies.[70] If such a tool does not exist on the market, the state itself provides one.[71] Currently, the so-called "Broadband Atlas"[72] is available on the website of the Federal Ministry of Transport.

The Broadband Atlas is designed as a central and free information medium that provides information on the current broadband coverage situation. The data is updated regularly. Interactive maps show not only the respective bandwidth but also the technologies available at the location. The information is displayed for each city district. Customers can also find out which providers are active locally and report deviating measurement results for internet speed, which will correct the display in the atlas if necessary. At breitbandmessung.de, a map with the results of all broadband measurements carried out to date is available. This allows customers to check measurement results for their region, which are in turn provided by individual providers. The results of the measurements are published by the Federal Network Agency in its annual report on broadband measurement.

[67] § 163 para. 6 TKG.

[68] Hierzu die Begründung des federführenden Ausschuss für Wirtschaft und Energie (2021), p. 468.

[69] Ausschuss für Wirtschaft und Energie (2021), p. 468.

[70] § 80 para. 4 TKG.

[71] § 80 para. 4 S. 1 TKG.

[72] Bundesministerium für Verkehr und digitale Infrastruktur (2021).

10 Independent Comparison Portals

With the instrument of independent comparison tools, the consumer is to be enabled to obtain and compare information on the contracts and the price of the service without being influenced by third parties. In comparison to the broadband atlas, which provides information about the supply by the network operators at the place of residence, the focus is, thus, on the concrete supply contracts offered and the fees to be paid for them.

In future, the Federal Network Agency will ensure that consumers have free access to at least one independent comparison tool.[73] Consumers should, thus, be enabled to compare different internet access services and publicly available interpersonal telecommunications services. The comparison tools must be operated independently of providers of the services to be compared, and disclose the owners and operators.[74] Comparison tools which comply with the statutory criteria[75] are certified by the Federal Network Agency or a company commissioned by it at the request of the provider.[76] In this way, independence is to be ensured.

The regulation is prompted by European requirements and is due to the fact that in practice existing comparison portals often do not show the required independence. Often there are links between the comparison portals and the providers of the services in question. For Germany, however, the legislator comes to the conclusion in the explanatory memorandum to the TKG that there are no links between the comparison portals existing on the internet and the providers of the services concerned, which could lead to a distortion of the results.[77] For Germany, it can generally be assumed that consumers have a sufficient selection of comparison tools at their disposal.[78]

11 Information on the Quality of Services

A decisive selection criterion besides the price is the quality of the services. The necessary expertise to assess the quality of internet access services is lacking in the majority of the population. In order to enable consumers to make an informed decision for this criterion as well when choosing their contracts, a number of obligations are imposed on the providers of internet access services in order to be able to guarantee transparency even before the contract is concluded. Statements on

[73] § 53 para. 1 TKG.
[74] § 53 para. 2 TKG.
[75] § 53 para. 2 TKG.
[76] § 53 para. 3 TKG.
[77] Bundesrat (2021), p. 334 mit Verweis auf Bundeskartellamt (2019), pp. 39–43.
[78] Bundesrat (2021), p. 334.

the quality of services must be meaningful and comparable in order to benefit the customer.

There are comprehensive publication obligations, especially in the area of broadband and mobile internet. Information that is particularly relevant for consumers regarding the quality of service offered must be published.[79]

The legislator understands quality of service to mean, in particular, information on the advertised download and upload speeds as well as information on any limitation of the volume or speed.[80] These criteria are largely based on the requirements of the European Directive on Electronic Communications (ECD), which the German legislator had to implement. For internet access services, specifications are made for latency, delay variation and packet loss.[81] For definitions and the respective measurement methods, the EECC refers to the recommendations of the International Telecommunication Union.[82] In order to achieve Europe-wide comparability, the national regulatory authorities are to specify the measurement methods to be used by the operators.[83] The European regulatory association BEREC, together with the European Commission, has therefore published a 40-page guideline paper on the relevant parameters for quality of service.[84] This comprehensively describes the measurement methods to be applied, the content and format of the published information and the quality certification mechanisms.[85] This is also intended to reduce compliance costs.[86] The Federal Network Agency will also be guided by these guidelines when specifying the TKG requirements in the area of customer protection. Comparability is thus largely ensured.

The Federal Network Agency also provides consumers with useful information and tools in its "Consumer Portal". In the area of "Internet and Telephone", the range of services is subdivided into the areas of changing providers and moving, contract, billing, supply and disruption and network neutrality, among others. The Federal Network Agency also refers to the respective rights of the consumer in the event of a detected poor performance by the service provider.

[79] § 52 para. 1 TKG.

[80] Bundesrat (2021), p. 333.

[81] Annex X ("Quality-of-Service Parameters, Definitions and Measurement Methods referred to in Article 104") of the EECC.

[82] International Telecommunication Union (n.d.).

[83] Article 104 para. 2 subpara. 1 EECC.

[84] Body of European Regulators for Electronic Communications (2020).

[85] Article 104 para. 2 subpara. 2 EECC.

[86] Recital 272 EECC.

12 Comprehensive Information on the Contents of a Contract

In addition to comparable information on the quality of services, the TKG imposes a number of other obligations on providers of internet access services in the context of contract initiation. In this way, customer protection is already to be implemented pre-contractually.[87] The transparency obligations intend to ensure that consumers have expert knowledge of the very complex matter before concluding a contract.[88] For customers, the offers become comparable, which in the end strengthens competition. In particular, in view of the technically charged and complex business conditions of the providers, it should first become comprehensible to them which services they are obliged to provide or which services they are promised.

Providers of internet access services must provide the information to be published in a clear, comprehensible and easily accessible manner in a machine-readable form.[89] The law contains a conclusive catalogue of information obligations that must be publicly accessible. The obligations affect all providers who make the provision of services subject to terms and conditions (T&Cs).[90] The information to be made public includes the applicable prices and tariffs, the duration of the contract and the charges applicable in the event of early termination of the contract, as well as rights regarding the termination of packages of offers or parts thereof, the standard terms and conditions for access to and use of the services they provide to end-users and consumers, the quality of service including an offer for verifiability of the data transmission rate, details of products and services specifically intended for users with disabilities and the actual, location-based mobile network coverage, including a map display of the current network coverage.

Providers of internet access services must provide further information: Company contact details, the scope of services offered and main features of each service provided, including the minimum level of quality and any usage restrictions imposed on terminal equipment provided, tariffs of the services offered, with details of the volume of communications included in specific tariffs and the applicable tariffs for additional communication units, numbers or services subject to special pricing conditions, access charges, maintenance charges, usage charges of any kind, special as well as target group-specific tariffs and additional charges as well as costs for terminal equipment, their general terms and conditions and the contract periods offered by them, the conditions for changing providers, termination conditions as well as procedures in connection with the transfer of telephone numbers or other identifiers, general and provider-related information on dispute resolution procedures and information on basic rights of end-users of Internet access services or

[87] See Bundesrat (2021), p. 470.

[88] Annex IX Directive (EU) 2018/1972.

[89] § 52 para. 3 sentence 1 TKG.

[90] § 52 para. 1 TKG.

publicly available interpersonal telecommunications services. This includes, in particular, itemised bills.

More specific information obligations can result from ordinances.[91] The Transparency Ordinance of the Federal Network Agency of 2017 deals, in particular, with internet access services.[92] In the current Transparency Ordinance, the providers of telecommunication services are obliged in § 1 to present the relevant information in so-called product information sheets in order to provide the consumer with a quick overview of the tariff details.[93] The Federal Network Agency provides standardised sample information sheets to ensure that the information is presented in a uniform and simple manner. In the future, it will also be possible to further specify the transparency requirements by ordinance.[94]

Effective consumer protection is thus to be created even before the conclusion of a contract by ensuring transparency.

13 Protection in the Event of Deviations from the Agreed Quality of Service

The TKG grants the consumer new rights in the event that the actual speed deviates from the speed indicated and guaranteed by the provider. In these cases, the consumer is entitled, without prejudice to other legal remedies, to reduce the contractually agreed charge for the part of the contract not provided in conformity with the contract or to terminate the contract extraordinarily. The right of reduction was only introduced by the TKG 2021.

The deviation complained of must have been determined by a monitoring mechanism provided by the Federal Network Agency or certified by it or a third party commissioned by it.[95] In the accompanying explanatory memorandum to the draft, the legislator refers for this purpose to the "measurement tool" of the Federal Network Agency presented at the beginning, such as the app "breitbandmessung".[96] The tool is available as a desktop application and as an app for mobile connections. Measurements are free of charge for everyone, regardless of provider and technology. Internet customers can, thus, independently check the performance of their broadband Internet access and compare the actual data transfer rate with the contractually agreed rate. This makes it possible to prove to the telecommunications

[91] For example, the ordinance on the promotion of transparency and cost control in the telecommunications market.

[92] Ordinance on the Promotion of Transparency in the Telecommunications Market of 19 December 2016 (TK-Transparenzverordnung).

[93] See for example Telekom Deutschland GmbH (n.d.).

[94] § 52 para. 4 TKG.

[95] § 57 para. 4 no. 1 TKG.

[96] Bundesrat (2021), p. 339.

provider that the service does not comply with the contract.[97] To do this, users must carry out a so-called measurement campaign.

However, the consumer can only assert his rights in the case of significant, continuous or regularly recurring deviations from the service stated by the provider. In this way, the legislator strikes an appropriate balance of interests between the consumer and the service provider. The Federal Network Agency has concretised the undefined legal terms of significant, continuous or regularly recurring deviation with a notice. Accordingly, the download is decisive. Poor performance exists if (1) 90% of the contractually agreed maximum speed is not achieved at least once on each of at least two measurement days, (2) the normally available speed is not achieved in 90% of the measurements or (3) the contractually agreed minimum speed is not achieved on each of at least two measurement days.[98]

With the newly granted rights, the legislator responds to the "multitude of customer complaints about" slow internet "and lack of fault clearance".[99] Until now, in cases where the actual speed differed from the speed stated by the service provider, consumers only had the option of agreeing on a contract adjustment with the provider or terminating the contract according to the provisions of the BGB.[100] If no agreement could be reached with the provider on a contract adjustment, the consumer had no possibility, with the exception of termination, to defend himself against the poor contractual performance.[101] Case law had recognised a special right of termination, in particular for cases in which the agreed speed was actually undercut by significantly more than half.[102]

14 Change After the end of the Contract Period

Notwithstanding the new options given to the consumer in the event of poor performance, the consumer is of course free to change providers after the contract period has expired. Consumer advocates regularly criticise the long contract periods of 24 months that are prevalent in the telecommunications sector. During this period, technology regularly develops at a considerable pace, so that the conditions for a certain service quickly become no longer "up-to-date". In the legislative process for the TKG amendment 2021, there were therefore efforts to limit the maximum contract term to 12 months, but this could not be pushed through in the parliamentary process. The contract term is still capped at a maximum of 24 months. What is new, however, is the right to terminate the contract with one month's notice if the contract

---

[97] For legal consequences according to TKMog see Sect. III. 3.

[98] Bundesnetzagentur (n.d.).

[99] Bundesrat (2021), p. 233.

[100] Bundesrat (2021), p. 339.

[101] Bundesrat (2021), p. 339.

[102] See only AG Fürth, judgment of 30 March 2009—340 C 3088/08 = NJOZ 2010, 616.

term is tacitly extended.[103] For this type of contract extension, the law now grants this special protection and thus introduces the "termination by mouse click".

Even if a customer decides to change their provider, the law protects the end-user's entitlement to internet access during this phase (change of provider procedure[104]). In particular, the receiving and the giving provider must ensure that there is no interruption of service during the phase in which the end user changes provider.[105] Providers must ensure that there is no interruption in the service provided by the supplying provider to the end-user before the conditions for switching are met.[106] For the days in which the transferring provider continues to provide its services under this obligation after the end of the contract, it can demand corresponding payment from the end user.[107] In addition, providers must advise end-users annually on the best tariff for them, taking into account the volume of services used by the end-user.[108]

Special case: Dealing with network congestion in times of pandemic

Due to the sharp increase in the use of data-intensive services as a result of the Corona pandemic, some providers have argued that they must throttle certain services, such as video streaming. Otherwise, network operation could no longer be maintained. These announcements have triggered controversial political debates due to the Net Neutrality Regulation (EU) No. 531/2012 (NN-V0). According to this, providers are to treat all traffic equally without discrimination, restriction or interference when providing internet access services. This obligation applies regardless of the sender and receiver, the content accessed or distributed, the applications or services used or provided, or the terminal equipment used.[109]

This principle can only be breached if efficient data traffic is otherwise not possible. The NN Regulation therefore does not prevent providers of internet access services from applying "reasonable" traffic management measures. Such measures are considered reasonable, if they are transparent, non-discriminatory and proportionate. They must not be based on commercial considerations. Nor may they use techniques designed to monitor specific traffic content. The measures shall not be maintained for longer than necessary.

In addition, the NN-VO names three constellations in which an exception to the prohibition of traffic management is permissible.[110] According to this, traffic management is permissible if, firstly, it is a matter of implementing government requirements (e.g. regulations for the protection of public safety, measures under criminal

[103] § 56 para. 3 TKG.

[104] Bundesrat (2021), p. 342.

[105] § 59 para. 1 sentence 4 TKG.

[106] § 59 para. 2 sentence 1 TKG.

[107] § 59 para. 3 TKG.

[108] § 57 para. 3 TKG.

[109] Art. 3 para. 3 NN-VO.

[110] Art. 3 para. 3 subpara. 3 der NN-VO.

law), secondly, the integrity and security of the network are maintained and, thirdly, an imminent network overload is prevented.

As stated by to the Federal Network Agency, a review of the providers' statements showed that the networks were stable.[111] According to the agency's assessment, malfunctions reported by internet users often have causes external to the network (e.g. insufficiently dimensioned capacities for home offices on the servers, equipment of the internet access points). It is already possible to reduce the traffic load if, for example, providers of video streaming services such as Netflix voluntarily feed in applications and content in lower quality—e.g. in SD/HD instead of UHD (4k). Internet access providers may apply traffic management measures (e.g. throttling) during the period of network congestion in relation to certain traffic categories (e.g. video streaming). Here, the entire data traffic category should be throttled equally and not only individual providers. Throttling video streaming could allow video conferencing services to be prioritised.

Due to the dynamics of the situation, the Federal Network Agency assumes that network congestion could occur in the future due to a further increase in the use of internet telephony, video conferencing and streaming services. It has therefore initiated a close dialogue with the providers to discuss possible measures to prevent network congestion. So far, it has not been necessary to take such measures.

15 Conclusion

The Corona pandemic abruptly demonstrated to politicians and administrators in Germany what backlogs existed in network expansion and the significance of a fast Internet access for teleworking and telelearning. Although these deficits had long been known in the circle of experts, it had not been feasible to provide the resources necessary for network expansion. Modern societies, apparently, have considerable difficulties in pursuing medium and long-term goals. Only when a crisis occurs can the necessary majorities in the economy and society be mobilised. This is not good news for tackling climate change. On the other hand, the extreme weather conditions in Germany are leading to a gradual awareness among the population of how important a responsible policy of sustainability is today. For the telecommunications sector, this creates new requirements. In addition to the goals of securing competition, consumer protection and infrastructure development, sustainability and the resilience of societal systems should gain in importance. They should be included in the catalogue of objectives of the TKG and the EECC. Empowering the consumer side to obtain the broadband services they need and to verify the fulfilment of providers' promises is a first step. In a second step, however, these specifications should be supplemented by the inclusion of sustainability criteria such as an energy balance of the products available on the market. Then these can also become relevant

[111] Bundesnetzagentur (2020).

as part of the decision for or against a contract offer. Of course, this does not require a comprehensive conception of how the goals of digitalisation, sustainability and resilience can be brought to bear. Their elaboration is a task for the near future. In this area, too, there is no way around the fact that the Federal Network Agency will drive and moderate this process.

References

Ausschuss für Wirtschaft und Energie (2021) Beschlussempfehlung und Bericht, BT-Drs. 19/28865

Body of European Regulators for Electronic Communications (2020) Guidelines detailing quality of service parameters. https://berec.europa.eu/eng/document_register/subject_matter/berec/regulatory_best_practices/guidelines/9043-berec-guidelines-detailing-quality-of-service-parameters

Body of European Regulators for Electronic Communications (2021) Summary report on BEREC sustainability ENG workshops: sustainability within the digital sector. What is the role of BEREC? https://www.berec.europa.eu/en/document-categories/berec/reports/summary-report-on-berec-sustainability-eng-workshops-sustainability-within-the-digital-sector-what-is-the-role-of-berec-0

Bundeskartellamt (2019) Sektoruntersuchung Vergleichsportale. April, 2019. https://www.bundeskartellamt.de/SharedDocs/Publikation/DE/Sektoruntersuchungen/Sektoruntersuchung_Vergleichsportale_Bericht.pdf?__blob=publicationFile&v=7

Bundesministerium für Verkehr und digitale Infrastruktur (2020) Aktuelle Breitbandverfügbarkeit in Deutschland (Stand Ende 2020). https://www.bmvi.de/SharedDocs/DE/Publikationen/DG/breitband-verfuegbarkeit-ende-2020.pdf?__blob=publicationFile

Bundesministerium für Verkehr und digitale Infrastruktur (2021) "Der Breitbandatlas". https://www.bmvi.de/DE/Themen/Digitales/Breitbandausbau/Breitbandatlas-Karte/start.html

Bundesnetzagentur (2017) Tätigkeitsbericht Telekommunikation 2016/2017. https://www.bundesnetzagentur.de/SharedDocs/Downloads/DE/Allgemeines/Bundesnetzagentur/Publikationen/Berichte/2017/TB_Telekommunikation20162017.pdf?__blob=publicationFile&v=3

Bundesnetzagentur (2019) Tätigkeitsbericht Telekommunikation 2018/2019. https://www.bundesnetzagentur.de/SharedDocs/Mediathek/Taetigkeitsberichte/2019/TK_20182019.pdf?__blob=publicationFile&v=9

Bundesnetzagentur (2020) Covid-19-Pandemie. Leitfaden zu Verkehrsmanagementmaßnahmen/sonstigen Maßnahmen im Falle von Netzüberlastungen vom March, 25, 2020. https://www.bundesnetzagentur.de/SharedDocs/Downloads/DE/Sachgebiete/Telekommunikation/Unternehmen_Institutionen/Netzneutralitaet/Corona/LeitfadenVerkehrsmanagementmassnahmen.pdf?__blob=publicationFile&v=3

Bundesnetzagentur (2021) Verbraucher-Radar Erstes Halbjahr 2021. https://www.bundesnetzagentur.de/DE/Vportal/Verbraucher-Radar/Radar_01_2021.pdf?__blob=publicationFile&v=2

Bundesnetzagentur (n.d.) Mitteilung Nr. 486/2017, Amtsblatt Nr. 13/2017 vom 12.07.2017 zur Konkretisierung der unbestimmten Rechtsbegriffe "erhebliche, kontinuierliche oder regelmäßig wiederkehrende Abweichung bei der Geschwindigkeit" bei Festnetz-Breitbandanschlüssen im Download gemäß Art. 4 Abs. 4 Verordnung (EU) 2015/2120 u. a. über Maßnahmen zum Zugang zum offenen Internet. https://www.bundesnetzagentur.de/SharedDocs/Downloads/DE/Sachgebiete/Telekommunikation/Unternehmen_Institutionen/Breitband/Breitbandgeschwindigkeiten/Zusammenfassung_und_Bewertung_der_Stellungnahmen.pdf?__blob=publicationFile&v=4

Bundesrat (2021) Gesetzentwurf der Bundesregierung. Drucksache 29/21

European Parliament (2021) Internet access as a fundamental right. https://www.europarl.europa.eu/RegData/etudes/STUD/2021/696170/EPRS_STU(2021)696170_EN.pdf

Freese B (2020) Mobilfunknetzausbau 5. Generation (5G) – Eine rechtliche Potentialanalyse. Netzwirtschaften und Rech 12(1):22–29

Holznagel B (2018) Zukunftsfähigkeit der Regulierungskonzeption. In: Holznagel (Hrsg) 20 Jahre Verantwortung für Netze. Bestandsaufnahme und Perspektiven. C.H. Beck, German

International Telecommunication Union (n.d.) Recommendation Y.2617: Quality of service guaranteed mechanisms and performance model for public packet telecommunication data networks. https://www.itu.int/rec/T-REC-Y.2617-201606-I/en

Rat für Nachhaltige Entwicklung (2021) Klimaneutralität. Optionen für eine ambitionierte Weichenstellung und Umsetzung – Positionspapier, Juni 2021. https://www.leopoldina.org/uploads/tx_leopublication/2021_RNE_Leopoldina_Klimaneutralitaet_geschuetzt.pdf

Säcker FJ, Körber T (2023) Kommentar TKG-TTDSG

Telekom Deutschland GmbH (n.d.) Produktinformationsblatt gem. § 1 TK-Transparenzverordnung. https://www.telekom.de/produktinformationsblatt/magentazuhause-xl.pdf

Wissenschaftlicher Beirat der Bundesregierung Globale Umweltveränderungen (2019) Unsere gemeinsame digitale Zukunft. Hauptgutachten. WBGU, Berlin. https://www.wbgu.de/fileadmin/user_upload/wbgu/publikationen/hauptgutachten/hg2019/pdf/wbgu_hg2019.pdf

The Right to Education in a Digital Era

Suzana Kraljić

Abstract The right to education is guaranteed in many international human rights documents. This right is especially important for children. Therefore, the States should protect, respect, and fulfil the children's right to education. Nowadays, also the right to education is facing the rapid development of technology. The COVID-19 outbreak additionality (re)opened questions about how to fulfil this child's right without discrimination or without widening the gap between children who have access to and use (good) digital environment and those without (good). In the article, particular emphasis will be given to the right to education in the digital especially from the present COVID-19 situation.

Keywords Right to education · Digitalization · Digital literacy · Digital divide · COVID-19

1 Introduction

Nowadays, technology is evolving and expanding faster than ever before. Following United National Secretary-General António Guterres (United Nations – Secretary General 2020),[1] managing rapid technological change is a defining challenge of our generation. The digital revolution affects now every country and territory and all areas of our private and professional lives. In our everyday lives, we face the necessities for (at least) the basic levels of digital competence. The internet has become a powerful and essential global public good that requires the highest possible level of international cooperation (United Nations – Secretary General 2020).[2]

[1] United Nations – Secretary General (2020).

[2] Ibid.

S. Kraljić (✉)
University of Maribor, Maribor, Slovenia
e-mail: suzana.kraljic@um.si

L. V. Sannikova (ed.), *Digital Technologies and Distributed Registries for Sustainable Development*, Law, Governance and Technology Series 64,
https://doi.org/10.1007/978-3-031-51067-0_8

Even our children are born and are growing with automation, robotization, digitalization, artificial intelligence, biometrics, virtual and augmented reality, which already has huge impact on their lives (Committee on Culture and Education 2018; Committee on the Rights of the Child 2021).[3] Access to and use the digital environment is essential for realizing children's rights and fundamental freedoms, including their education, health, participation, and maintaining family and social relationships. Where children do not have access to the digital environment or where this access is limited as a result of poor connectivity, their ability to fully exercise their human (also children) rights may be affected (par. 10 of the Recommendation CM/Rec(2018)7 of the Committee of Ministers to member States on Guidelines to respect, protect and fulfil the rights of the child in the digital environment).

Digitalization is also influencing education. The educational systems are facing rapid technological changes on the one hand. Still, on the other hand, the education systems are struggling to adapt, not only in terms of how to use the technologies but also in teaching the skills and training the teachers and learners (Committee on Culture and Education 2018).

Especially, in the last 36 months, since the COVID-19 outbreak, the educational systems on one hand and teachers and learners on the other hand, have been faced with enormous changes in teaching methods. The schools (from primary to universities) have made a quick transition from regular to on-line classes. After 36 months, we are still not at the end of the corridor. We can already say that COVID-19 brought a significant reform in our education systems and is still not finished. The new teaching methods are planned, the acceleration of digitalization (for example, hybrid classes for this and potential future emergencies, new teaching tools such as MsTeams, Webex, Zoom, Skype, Messenger and others).

The article will present the right to education as a children's right. Particular emphasis will be given to the right to education in the digital, especially from the present COVID-19 situation.

2 General on Right to Education

In 1989, the United Nations adopted the Convention on the Rights of the Child (hereinafter: CRC). CRC is today the most widely ratified human rights treaty in history. All states except the United States ratify it. CRC is the first international binding instrument that explicitly recognized children as human beings with innate rights (Odink 2019, p. 2).[4] CRC has become the landmark treaty on children's rights, outlining universal standards for all children's care, treatment, survival, development, protection, and participation. CRC recognizes the right of all children to have the best possible start in life, grow up happy and healthy, and develop to their full

[3] Committee on Culture and Education (2018) and Committee on the Rights of the Child (2021).

[4] Odink (2019).

potential (European Commission 2021a).[5] The goal of education is the development of the child's personality, talents, and mental and physical abilities to their fullest potential and the preparation of the child for a responsible life in a free society (par. 40 of the Recommendation CM/Rec(2018)7 of the Committee of Ministers to member States on Guidelines to respect, protect and fulfill the rights of the child in the digital environment). In support of this goal, the knowledge and resources of the digital environment must be available to all children from the earliest years. The development of the digital literacy (also so-called digital citizenship education) must be provided in a way that is inclusive and takes into account children's evolving capacities and the particular circumstances of children in vulnerable situations.

Following the Committee on the Rights of the Child (2021), the rights of every child must be respected, protected and fulfilled also in the digital environment. Innovations in digital technologies affect children's lives and their rights in wide-ranging and interdependent ways, even where children do not themselves access the internet. Meaningful access to digital technologies can support children to realize the full range of their civil, political, cultural, economic, and social rights. The digital environment has opened up many new opportunities and challenges for children. Children play, create, learn, interact and express themselves in an on-line and connected environment from a very young age. Digital technologies make it possible to children to be part of global movements and play the role of active citizens. As digital natives, they are better placed to thrive in an increasingly digitalized and connected education and future labor market systems (European Commission 2021a).

CRC sets out the children's rights across all areas of their lives and one of these areas represents education (Kraljić 2020).[6] Article 28 of the CRC gives the basis to the child's right to education:

1. States Parties recognize the right of the child to education, and with a view to achieving this right progressively and on the basis of equal opportunity, they shall, in particular:
    (a) Make primary education compulsory and available free to all;
    (b) Encourage the development of different forms of secondary education, including general and vocational education, make them available and accessible to every child, and take appropriate measures such as the introduction of free education and offering financial assistance in case of need;
    (c) Make higher education accessible to all on the basis of capacity by every appropriate means;
    (d) Make educational and vocational information and guidance available and accessible to all children;
    (e) Take measures to encourage regular attendance at schools and the reduction of drop-out rates.
2. States Parties shall take all appropriate measures to ensure that school discipline is administered in a manner consistent with the child's human dignity and in conformity with the present Convention.

[5] European Commission (2021a).

[6] Kraljić (2020), pp. 27–44.

3. States Parties shall promote and encourage international cooperation in matters relating to education, in particular with a view to contributing to the elimination of ignorance and illiteracy throughout the world and facilitating access to scientific and technical knowledge and modern teaching methods. In this regard, particular account shall be taken of the needs of developing countries.

Concerning Article 28 of the CRC, every child should enjoy the right to education. But despite that we are living in the twenty-first century, today still many children have limited access to the right to education. The limitations could be based on different circumstances caused by humans (for example, wars, discriminations) or nature (for example, natural disasters, epidemics such as COVID-19). Also, the General Comment No. 14 of the Committee on the Rights of the Children provides that it is in the children's best interests to have access to quality education, including early childhood education, non-formal or informal education, and related activities free of charge. All decisions on measures and actions concerning a specific child or a group of children must respect the best interests of the child or children, with regard to education (Committee on the Rights of the Children 2013, p. 17).

Today the right to education is highlighted as a common good, one of the most important human rights, and a special right of the child. Education is both a human right and an indispensable means of realizing other human and child rights (Committee on Economic, Social and Cultural Rights 1999).[7] It is a crucial creator of a child's life before and after they reach full age. Primary education is seen as the 'passport to life' (Tawil and Cougoureux 2013, p. 7).[8] Namely, education can not be seen just as an investment in the child's future but also as an opportunity for joyful activities, respect, participation, and fulfillment of ambitions (Committee on the Rights of the Children 2013, p. 17).

From the education can benefit children, children's family and also broader society. Education is powerful. It can break generational cycles of poverty by enabling children to gain the life skills and necessary knowledge needed to cope with today's and future challenges. Education is also strongly linked to concrete improvements in nutrition and health, improving children's chances for survival. Education brings children the empowerment to be full and active participants in their present and future society, able to exercise their rights and duties and engage in civil and political life. Education is also a powerful protection factor: children who are in school are less likely to come into conflict with the law and much less vulnerable to rampant forms of child exploitation, including child labor, trafficking, and recruitment into armed groups and forces (Human Rights Watch 2016, p. 2).[9] The latter Human Rights Watch (2016, p. 3), particularly highlights the 'lost generation' of children from Syria, 2.1 million of whom do not go to school in Syria, and nearly 1 million who live in neighboring countries as refugees.

[7]Committee on Economic, Social and Cultural Rights (1999).

[8]Tawil and Cougoureux (2013).

[9]Human Rights Watch (2016).

Following the idea of Katarina Tomaševski,[10] education should be based on the so-called "4-As scheme". The governments should make education available, accessible, acceptable, and adaptable:

(a) *availability*—that education should be free and government-funded and that there should be adequate infrastructure and trained teachers, who should be able to support education delivery (for example, safe building, enough teachers, libraries, computers, sanitation facilities, books);
(b) *accessibility*—the system must be non-discriminatory (for example, especially for vulnerable groups as Roma, indigenous, disabled children, refugees, internally displaced children, stateless children) and accessible to all (for example, transport, safe way to school). The states should guarantee especially the inclusion of the most marginalized. The states eliminate legal and administrative barriers and obstacles to compulsory schooling (such as fees, distance, schedule);
(c) *acceptability*—the educational content must be relevant, non-discriminatory, and culturally appropriate, and of quality. The school itself should be safe, and teachers should be professional. Acceptability refers to both learners and parents, in terms of content and teaching methods. Quality refers at least to the provision of minimum educational standards (for example, safety, environmental health);
(d) *adaptability*—education can evolve with the changing needs of society and contribute to challenging inequalities, such as gender discrimination, and that it can be adapted locally to suit specific contexts. The adaptability of education is reflected in adapting to the individual and the narrower and broader society. The adaptation addresses, in particular, the needs of children who come from specific social and cultural backgrounds (for example, minority and indigenous children, working children, children with disabilities, child migrants) (Tomaševski 2009, pp. 12–14; Hajrullaia and Saliua 2016; Šimenc and Kodelja 2015, p. 210).[11] Additionally, Bucataru (2016)[12] particularly highlights transgender children, who also suffer from discrimination and their right to education is often not protected.

The idea of "4-As scheme" of Katarina Tomaševski is also incorporated in the European Parliament Resolution of 12 June 2018 on modernisation of education in the EU (*whereas the right to education is a fundamental human right and education in all its forms and at all levels must exhibit the following interrelated and essential features: a) availability; b) accessibility; c) acceptability; and d) adaptability*) (par.

[10] Katarina Tomaševski was from 1998 to 2004 the first United Nations Special Rapporteur on the right to education of the Commission on Human Rights. She was also known as a defender of education as a human right (see more Klees and Thapliyal 2007, pp. 497–510; Monteiro 2010, 1988–1992). Later this function was performed by Vernor Muñoz Villalobos (Costa Rica, 2004–2010), Kishore Singh (India, 2010–2016), Koumnou Boly Barry (Burkina Faso, since 2016).

[11] Tomaševski (2009); Hajrullaia and Saliua (2016), pp. 70–74; Šimenc and Kodelja (2015), pp. 195–215.

[12] Bucataru (2016), pp. 59–81.

B). The Resolution on modernisation of education in the EU affirms that universal quality education is an essential component of personal, cultural, social, and professional development in a knowledge-based society (par. 1). The Resolution on modernisation of education in the EU also addresses digitalization. In par. 15 is explicitly written, "*that digitalization and the establishment of common educational platforms offer for modern education, especially in terms of distance learning, distance education, and blended learning, which should allow more flexibility in education by tailoring it more closely to learners' individual living situations and thereby benefiting lifelong learning, education quality, accessibility and the development of future skills; highlights the need for age-appropriate ICT and media curricula that respect child development and wellbeing and emphasise the importance of both responsible use and critical thinking.*"

The State has an essential role in ensuring that the right to education is respected, fulfilled, and protected. The obligation to respect requires States to avoid measures that would hinder or prevent the enjoyment of the right to education. The obligation to protect requires States to take measures that prevent third parties from interfering with the enjoyment of the right to education. The obligation to fulfill obliges States to adopt positive measures that enable and assist individuals and communities in enjoying the right to education. The State must act as a guarantor of the right to education (UNESCO 2015, p. 75).[13]

Unfortunately, practice is quite different from that required by rights law pertaining to humans and children (Klees and Thapliyal 2007, pp. 502–503).[14] Katarina Tomaševski stated in 2006 that an increase in school enrolments from 40% to 60% could be seen as a huge success, but we should not forget that 40% of the children are still excluded from school. Therefore, 40% of children still can not fulfill their right to education as a violation (Human Rights Watch 2016, p. 1). Various studies (Kaur and Singh 2014; UNDP Bangladesh n.d.; Zhang et al. 2015)[15] have found significant differences in education between rural and urban areas. According to the Food and Agriculture Organisation of the United Nations (FAO), in 2015, about 57 million primary-school-age children did not attend school. Over four out of five of these children live in rural areas (FAO 2020).[16] This difference in knowledge and education levels between rural and urban areas is called the 'urban–rural education divide'. It is the main barrier to achieving universal primary education and is seen as a key obstacle to eradicating hunger and poverty, promoting gender equality, and empowering women (FAO n.d.).[17]

---

[13] UNESCO (2015).

[14] Klees and Thapliyal (2007), pp. 497–510.

[15] Kaur and Singh (2014), pp. 125–132; UNDP Bangladesh (n.d.); Zhang et al. (2015), pp. 196–224.

[16] FAO (Food and Agriculture Organization of the United Nations) (2020).

[17] FAO (Food and Agriculture Organization of the United Nations) (n.d.).

3 Digital Literacy

Following UNESCO (2015, p. 49),[18] the classroom-centered learning is nowadays challenged by expanding access to knowledge and the emergence of learning spaces beyond the traditional classroom. We live in a time of the fourth industrial revolution that involves a systemic change across many sectors and aspects of human life (Schwab 2017).[19] The fourth revolution has brought changes in work, communication, information, leisure, transport, etc. (Schwab 2017).[20] Digital technology is central to almost every aspect of the response to the pandemic, from vaccine research to on-line learning models and tools that enable hundreds of millions of people to work and learners to learn or study from home (United Nations – Secretary General 2020). Digitalization marks a new era that builds and expands the impact of digitization in new and unanticipated dimensions of use and dissemination. Its technologies are already changing all areas of our daily lives and as well as education.

Digital literacy is essential for life in a digitalized world (European Commission 2020). Digital literacies are those capabilities that fit an individual for living, learning and working in a digital society (UNICEF 2019, p. 9).[21] In addition to the term digital literacy, two other terms are describing the same content, namely 'digital competence' and 'digital citizenship'. Following European Commission, the term "*digital competence involves the confident, critical and responsible use of, and engagement with, digital technologies for learning, at work, and for participation in society. It includes information and data literacy, communication and collaboration, media literacy, digital content creation (including programming), safety (including digital well-being and competences related to cybersecurity), intellectual property-related questions, problem-solving and critical thinking*" (European Parliament 2018).[22] The concept of competence is understood as a combination of knowledge, skills, and attitudes. Including a distinction between these three dimensions is particularly important for children, who might have the skills to complete a certain digital task but might lack knowledge about the context and critical approach to performing that task. Also, such a categorization can help adapt digital literacy frameworks into educational curriculums, which are generally based on knowledge, skills, and attitudes (UNICEF 2019, p. 11).

Following the Council of Europe (hereinafter: CoE), that uses the term digital citizenship, which refers "*to the competent and positive engagement with digital technologies and data (creating, publishing, working, sharing, socializing, investigating, playing, communicating and learning); participating actively and responsibly (values, skills, attitudes, knowledge and critical understanding) in communities*

---

[18] UNESCO (2015).

[19] Schwab (2017).

[20] Ibid.

[21] UNICEF (2019).

[22] European Parliament (2018).

(local, national, global) at all levels (political, economic, social, cultural and intercultural); being involved in a double process of lifelong learning (in formal, informal, non-formal settings) and continuously defending human dignity and all attendant human rights" (UNICEF 2019, p. 11).

The CoE is probably the most advanced international organization in terms of policymaking in the field of children and the digital society, under the so-called 'digital citizenship education'. Recognizing that digital citizenship is essential for its three missions related to human rights, democracy and the rule of law, the CoE has in 2018 adopted already mentioned 'Guidelines to Respect, Protect and Fulfil the Rights of the Child in the Digital Environment' and the 'Strategy for the Rights of the Child' (2016–2021). Guidelines to Respect, Protect and Fulfil the Rights of the Child in the Digital Environment and Strategy for the Rights of the Child form a solid set of binding rules to assist states in developing policies for safeguarding children's interests in digital environments (UNICEF 2019, p. 15).

4 From Digital Divide to Digital Chasm?

The term 'digital divide' was introduced in the middle 1990s. Traditionally, the digital divide was referred to those individuals and communities having access and those without access. Following DiMaggio and Hargittai (2001, p. 2),[23] the term digital divide refers to the gap of those who are to benefit from the digital age and those who do not. With the development of the digitalization, the definition has become more concrete.

International conventions and constitutions[24] imply that education shall be free, at least in the elementary and fundamental stages.[25] Because of the COVID-19

[23] DiMaggio and Hargittai (2001).

[24] For example, constitutions of Slovenia (Article 57 and Article 52(2)); Croatia (Article 66); Egypt (Article 19); Latvia (Article 122); Montenegro (Article 75 and Article 79(4)); Norway (Article 109); Russia (Article 43); Tajikistan (Article 41); Ukraine (Article 53) and others.

[25] See for example: Article 26 of the UNDHR (1948 – '*Education shall be free, at least in the elementary and fundamental stages. Elementary education shall be compulsory.*'); UNESCO CADE (1960 – '*The States Parties to this Convention undertake to formulate, develop and apply a national policy which, . . . will tend to promote equality of opportunity and of treatment . . . and in particular: (a) To make primary education free and compulsory.*'); Article 13 ICESCR (1966 – '*Primary education shall be compulsory and available free for all.*'); Article 2 of the Protocol No. 1 of the ECHR ('*No person shall be denied the right to education.*'); Article 14 CFREU (2012 – '*1. Everyone has the right to education and to have access to vocational and continuing training. 2. This right includes the possibility to receive free compulsory education. 3. The freedom to found educational establishments with due respect for democratic principles and the right of parents to ensure the education and teaching of their children in conformity with their religious, philosophical and pedagogical convictions shall be respected, in accordance with the national laws governing the exercise of such freedom and right.*'); European Social Charter (1996 revised – '*With a view to ensuring the effective exercise of the right of children and young persons to grow up in an environment which encourages the full development of their personality and of their physical and*

pandemic, the digital infrastructure became a prerequisite for the fulfillment of this right. Therefore, the states should have an obligation to ensure the appropriate digital infrastructure. Its delivery should be universal, but in practice, for example, the cost of on-line access falls on families privately (Digital Freedom Fund 2020).[26]

Nowadays, the digital divide is more than ever an important public policy debate that encompasses social, economic and political issues, and especially in education, which will be given particular emphasis in the present paper.

5 The Impact of COVID-19 on Education

The COVID-19 outbreak has impacted the daily functioning of educational systems at all levels (primary, secondary and high school/university). For most educational systems and institutions, this was a new situation. But, for example, the Cambridge University closed its doors for the second time in its history. The Cambridge University closed its doors for first time in 1665 due to a great black plague epidemic, caused by the bacterium *Yersenia Pestis,* that struck England (IESALC 2020, p. 9).[27] At that time Isaac Newton just finished his bachelor's degree at Cambridge University's Trinity College. Because of the school closure he returned to his home, Woolsthorpe Manor. This time is known as his 'year of wonders', because of the amazing breakthroughs and insights he made during that time, which laid the groundwork for modern physics. After the great plague ended (1667), he returned back to Trinity College and published his 'Mathematical Principles of Natural Philosophy', also known as the 'Principia' (Lopez-Guzman 2020).[28]

The digital divide in education was present even before the COVID-19 outbreak. Learners without computer ownership, high-speed access to the Internet, digital literacy and other information and communication technologies are put at a disadvantage in their education (Sanders and Scanlon 2021, p. 130).[29] The gap between those with sufficient knowledge of and access to technology and those without becomes with the COVID-19 even more extensive and deeper—the gap becomes chasm. In every class we can find learner(s) who has/have multiple or the most up-to-date laptop or mobile phone and has/have access to high-speed broadband. These learners are likely to reach better results and educational success than those who

mental capacities, the Parties undertake, either directly or in co-operation with public and private organizations, to take all appropriate and necessary measures designed: 2. to provide to children and young persons a free primary and secondary education as well as to encourage regular attendance at schools.').

[26] Digital Freedom Fund (2020).

[27] Institute for Higher Education in Latin America and the Caribbean (IESALC) (2020).

[28] Lopez-Guzman (2020).

[29] Sanders and Scanlon (2021), pp. 130–143.

have just one computer, which they must share with other family members (for example, siblings), and only have dial-up access to internet.

Due to the outbreak of the COVID-19 pandemic, the educational process shifted overnight in Slovenia and other states from 'the bench behind the screens' (Kraljić 2020, p. 38).[30] António Guterres, the 9th Secretary-General of the UN, remarks that the digital divide could be in present times of COVID-19 pandemia a matter of life and death for people who cannot access essential health-care information. It is threatening to become the new face of inequality, reinforcing the social and economic disadvantages in access to education (United Nations – Secretary General 2020).

Despite the current situation, which opened many new questions in national and global education systems, all necessary efforts should be made to ensure that everybody has access to digital education. Education is a fundamental human and especially children's right. Access to this right has to be guaranteed, independent of the environment in which it occurs: physical, digital, or both (European Commission 2020).

Following Skulmowski and Rey (2020),[31] the sudden and unexpected outbreak of COVID-19 required quick adaption to a new situation. The transformation of learning and teaching was for most schools (primary, secondary and universities) rapidly. They transformed the educational process into a digital format from one day to others. The COVID-19 pandemic highlighted the importance of digitization for European society. Digital technologies brought new ways to learn, entertain, work, explore, and fulfill ambitions. They also brought new freedoms and rights and allowed EU citizens to reach out beyond physical communities, geographical locations, and social positions.

The COVID-19 pandemic has demonstrated the need for fit education and training system for the digital age is essential. The new Action Plan was adopted and presented the vision for improving digital literacy, skills, and capacity at all levels of education and training and for all levels of digital skills (from basic to advanced) (European Commission 2020). Therefore, the new Digital Education Plan for the period 2021–2027 puts forward two priorities, covering together 13 actions in total:

(a) priority 1: fostering the development of a high-performing digital education ecosystem:

  aa) Action 1: strategic Dialogue with Member States on the enabling factors for successful digital education;
  bb) Action 2: Council Recommendation on blended learning for primary and secondary education;
  cc) Action 3: European Digital Education Content Framework;
  dd) Action 4: connectivity and digital equipment for education;

[30] Kraljić (2020), pp. 27–44.

[31] Skulmowski and Rey (2020), pp. 212–216.

ee) Action 5: digital transformation plans for education and training institutions;
ff) Action 6: artificial intelligence and data usage in education and training;

(b) priority 2: enhancing digital skills and competences for the digital transformation:

aa) Action 7: common guidelines for teachers and educators to foster digital literacy and tackle disinformation through education and training;
bb) Action 8: update the European Digital Competence Framework to include AI and data-related skills;
cc) Action 9: European Digital Skills Certificate (EDSC);
dd) Action 10: Council recommendation on improving the provision of digital skills in education and training;
ee) Action 11: cross-national collection of data on student digital skills and introduce an EU target for student digital competence;
ff) Action 12: digital Opportunity Traineeships;
gg) Action 13: women's participation in STEM (European Commission 2021b).

Because of the COVID-19 outbreak in 2020, most countries have closed their schools at least partly, if not entirely. As of 27 July 2020, schools in 107 countries had shut their doors (country-wide closures). The closures affected 1,066,817,855 learners or 60.9% of all enrolled learners. Still, the peak was reached on 2 April 2020 when country-wide closures were seen in 194 countries, impacting 1,598,099,008 learners or 91.3% of all enrolled learners (UNESCO 2020a).[32] According to the UNESCO – Institute for Statistics (2021), school closures caused by COVID-19 differs by country. For example: from 2020 (Q1) to 2021 (Q2), schools in Slovenia and Hungary had been fully closed for 21 weeks, in Romania for 22 weeks, in India for 25 weeks, in Serbia and Angola for 28 weeks, and in Brazil for 38 weeks.

On the other hand, Sweden, Australia, Russia, and Belarus have not had complete closure of schools at all. The most extended school closure was in Uganda, where it lasted for 43 weeks, and in Mexico, schools in some regions closed for 53 weeks. For comparison, the schools for the H1N1 pandemic in Mexico have been closed for 14 days, from 27 April to 10 May (Smith 2021).[33] In Sierra Leone, the schools remain closed for 11 weeks because of the COVID-19 outbreak, but in the Ebola outbreak for 9 months (Smith 2021).[34]

The COVID-19 and the school affected the amount of school work. For example, in Slovenia, respondents in research that involved ten European countries (Slovenia, Germany, Ireland, Romania, France, Austria, Spain, Italy, Norway, Portugal, Switzerland) said they had been faced with 63% more workload. In Portugal, the increased workload was 51% in Austria, France, Germany, Romania and Spain,

[32] UNESCO (2020a).

[33] Smith (2021), pp. 53–78.

[34] Ibid.

between 30% and 45% of children reported a similarly increased workload. On the other hand, in Ireland, Italy, and Norway, about half of the children said they had less workload (Vuorikari et al. 2020, p. 8).[35]

These new circumstances for only a few months have, according to initial estimates, already led to an "educational crisis" (Human Rights Council 2020, p. 1; Kraljić 2020). Following Giannini et al. (2021),[36] school closures and the resulting disruptions to school participation and learning are projected to amount to losses valued at $ 10 trillion in terms of affected children's future earnings.

In principle, the move itself to education to on-line distance learning was not a problem for many children, given they are very familiar with digitalization. Children very quickly gained new digital competencies in using digital technologies for on-line school activities. Following the abovementioned research, 69% of children in Slovenia (the highest increase was gained in Romania (72%) and lowest in Austria (50%)) gained in autonomy in using digital technologies, which enabled them to conduct the new method of schooling (teaching and learning). The level of digital competence of children has increased (Vuorikari et al. 2020, p. 8).

Yet, it should not be overlooked that the social and economic differences between children have become even more noticeable. Many children did not have a suitable computer, printer, or Internet connection to be included equally in this new way of teaching. It should also not be forgotten that many children did not have a private and quiet corner to permit them to integrate seamlessly into the pedagogical process and learning (Kraljić 2020).

As one of the most vulnerable groups, children are currently (in the 'COVID-19 crisis') further exposed to encroachments on their rights. Interventions may relate to restrictions or even violations. Although the circumstances arguably dictate justified restrictions on children's rights, special care must be taken to avoid violations and the resulting short- or even long-term consequences. A fundamental right of children currently in the spotlight is undoubtedly the right to education. Even though the implementation and continuation of distance learning were quickly organized and arranged, it should not be overlooked that this method is particularly susceptible to exacerbating the differences already present in the 'regular teaching process'. Namely, distance on-line teaching depends on modern technology (for example computer, telephone, a good Internet connection), which unfortunately could not be provided to all children. Half the total number of learners (around 826 million learners) kept out of the classroom by the COVID-19 pandemic do not have access to a household computer and 43% (706 million) have no internet at home (UNESCO 2020b).[37] This certainly infringes on children's right to have access to adequate education (Kraljić 2020).

Teaching and learning have become more informal, mobile and impersonal with the transition to 'behind the screen'. How deeply this has exacerbated the already

[35] Vuorikari et al. (2020).

[36] Giannini et al. (2021).

[37] UNESCO (2020b).

existing inequality (for example, between poor and rich, rural and urban, boys and girls) and how children's right to education has suffered may never be fully known as much remains hidden behind closed family doors (Kraljić 2020). For example, drop-out is on the rise (Dwawan 2020).[38] Already before the COVID-19 outbreak, dropping-out rates have been 66% higher for children living in stressful household conditions, deriving from families with low socio-economic status, with parents unemployed or/and uneducated, Roma, refugees or immigrants. Many children with disabilities also leave school earlier and just fewer learners with disabilities complete a university degree (14.4%) (European Commission 2021a, p. 9). These children are more likely to risk school failure and educational and social exclusion (Tsolou et al. 2021, p. 537).[39] For example, after the Ebola crisis and reopening the schools in Sierra Leone, girls were 16% less likely to be in school. Higher drop-out was likely accompanied by increased child labor and child marriage for children and adolescents (World Bank Group 2020). The parameters mentioned above have become more apparent in light of the new educational conditions that have occurred due to the COVID-19 pandemic and can also lead to named consequences.

Additional, to drop-out problem, we should also mention the so-called 'aged out' situation. Following Article 28(1)(a), '*states recognize the right of the child to education, and with a view to achieving this right progressively and on the basis of equal opportunity, they shall, in particular make primary education compulsory and available free to all*'. COVID-19 pandemic has led to school closure (full or partially) in many states worldwide. Meanwhile, many children won't be able to complete primary (or secondary) education because they have 'aged out' of free education during COVID-19 pandemic related school closures. These children likely did not fully benefit from the allotment of free education they were entitled to under law (Human Rights Watch 2021).

Therefore, it is crucial that on-line distance learning is only seen as a temporary solution for addressing the COVID-19 crisis. The digitalization of education should never replace on-site schooling with teachers (Human Rights Council 2020, p. 2).[40] 'Traditional' on-site learning can better focus on the needs of an individual child or a vulnerable group of children. The direct contact between teacher and learners makes it easier to identify the child's requirements and the possibilities for providing them. This also ensures a better exercise of the right to education and other children's rights (Kraljić 2020).

The COVID-19 pandemic significantly increased also the time children spend on-line—as well for educational purpose and leisure time. On the other hand, measures such as on-line teaching, movement, and contact restrictions (for example, curfews, social distancing) increased in the amount of time children spend with the computer or mobile phones. Therefore it is necessary to create proper legal instruments to protect children and their right in digital environment. EU already has

---

[38]Dwawan (2020), pp. 5–22.

[39]Tsolou et al. (2021), pp. 529–544.

[40]Human Rights Council (2020).

developed such legal instruments and policy initiatives which should ensure such protection—for example:

(a) Directive 2011/93/EU of the European Parliament and of the Council of 13 December 2011 on combating the sexual abuse and sexual exploitation of children and child pornography, and replacing Council Framework Decision 2004/68/JHA (OJ L 335, 17.12.2011, pp. 1–14);
(b) Council Framework Decision 2008/913/JHA of 28 November 2008 on combating certain forms and expressions of racism and xenophobia by means of criminal law *(OJ* L 328, 6.12.2008, pp. 55–58)*;*
(c) Directive 2010/13/EU on audiovisual media services (also known as Audiovisual Media Services Directive) (OJ L 95, 15.4.2010, pp. 1–24);
(d) Regulation (EU) 2016/679 of the European Parliament and of the Council of 27 April 2016 on the protection of natural persons with regard to the processing of personal data and on the free movement of such data, and repealing Directive 95/46/EC (also known as General Data Protection Regulation or shortly GDPR) (OJ L 119, 4.5.2016, pp. 1–99);
(e) Directive 2005/29/EC of the European Parliament and of the Council of 11 May 2005 concerning unfair business-to-consumer commercial practices in the internal market and amending Council Directive 84/450/EEC, Directives 97/7/EC, 98/27/EC and 2002/65/EC of the European Parliament and of the Council and Regulation (EC) No 2006/2004 of the European Parliament and of the Council ('Unfair Commercial Practices Directive') (Text with EEA relevance) (OJ L 149, 11.6.2005, pp. 22–39) *(*also known Directive on Unfair Commercial Practices);
(f) Communication on Tackling online disinformation: a European Approach, COM(2018)236;
(g) European Strategy for a Better Internet for Children, COM(2012)196.

Also the Council of Europe developed standards and adopted various international instruments which give basis for protection of child's best interests in the digital environment. For example:

(a) Council of Europe Guidelines to respect, protect and fulfil the rights of the child in the digital environment (also known as Child Rights Digital Guidelines);
(b) Convention for the Protection of individuals with regard to automatic processing of personal data;
(c) Convention on Cybercrime (also known as Budapest Convention);
(d) Convention on Protection of children against sexual exploitation and sexual abuse (also known as Lanzarote Convention);
(e) Convention on Preventing and combating violence against women and domestic violence (also known as "Istanbul Convention") etc.

Technology and digitalization can also lead to misuses, such as cyberbullying, criminal activity (for example, child pornography), and terrorism. Therefore, the learners from early years must be informed and taught how to face such risks and hazards. Namely, on-line learning also has a significant impact on children's rights and freedoms. Economic exploitation of their personal data, products with intrusive advertising, invasive behavioral surveillance or discrimination by-design are routine. These consequences reach beyond the school grounds and into private and family life (Digital Freedom Fund 2020).[41]

Children should be protected from the potentially harmful effects of on-line content and the risk of sexual abuse or exploitation. But also less serious risks should not be ignored (for example, advertisements, spam, disclosure of personal information, aggressive, violent, hateful pornographic, unwelcome and racist content). Following the abovementioned, increasing the time that children spent on-line, also led to an increase in on-line child sexual abuse. Due to their lack of critical thinking and experience, children are often unable to assess the long-term consequences of unwise on-line behavior. Following EUROPOL (2020, p. 17),[42] on-line child sexual abuse will remain a significant threat as long as children spend large amounts of time on-line unsupervised, either during their leisure time or while receiving education via remote learning arrangements. In order to prevent or at least limit such harmful effects and risks, the states must adopt and implement measures that should protect children from on-line harassment (for example for sexual purposes) and ensure that children will not become involved in illegal activities, financial scams, or terrorism (United Nations – Generally Assembly 2016, p. 13).

The already mentioned new Digital Education Action Plan (2021–2027) promotes digital literacy because of tackling disinformation and puts education and training at the heart of this effort. Internationally, guidance has just been released on interpreting the child's rights in the digital environment (Committee on the Rights of the Child 2021).

The imposition of on-line learning, which is possible only using digital technology, has also undermined parents' right to autonomy, who had chosen no-tech or low tech teaching for their children (Digital Freedom Fund 2020). Many parents have been obliged to follow and supervise their children's learning process during the COVID-19 school closure. On the one hand, they have been 'forced' to learn the school material again to better support their children. On the other hand, they also become aware of the complexity of teachers' work (UNESCO 2020c, p. 13).[43]

6 Right to Technology as a New Human Right?

The rapid development of digitalization is intertwined with the development of technology. Both have already influenced human rights and opened the question of whether we could speak about a new human right—the right to technology. We

[41] Digital Freedom Fund (2020).

[42] European Union Agency for Law Enforcement Cooperation (EUROPOL) (2020).

[43] UNESCO (2020c).

can find the basis for this new right already in Article 27(1) of the Universal Declaration of Human Rights defines that *'everyone has the right ... to share in scientific advancement and its benefits.'* Also, the International Covenant on Economic, Social and Cultural Rights in Article 15(1)(b) states that *'The States Parties to the present Covenant recognize the right of everyone:...(b) To enjoy the benefits of scientific progress and its applications'*. CRC in Article 28(3) also refers to the right to technology: '*States Parties shall promote and encourage international cooperation in matters relating to education, in particular with a view to contributing to the elimination of ignorance and illiteracy throughout the world and facilitating access to scientific and technical knowledge and modern teaching methods. In this regard, particular account shall be taken of the needs of developing countries.*' Following Sun (2020, p. 85),[44] the new fundamental right to technology has two major functions. The right to technology entitles everyone to enjoy the benefits of fundamental technologies and creates a mandate for ensuring the equal distribution of such benefits. On the other hand, Sun also states that the fundamental right to technology protects everyone from seriously harmful uses of all technologies. For example, when governments use digital technologies to unjustifiably or even shut down the internet, they undermine or completely deny individual freedom of information and the power of the internet to promote the socially beneficial exchange of information (Sun 2020, p. 86).[45]

7 Conclusion

Digitalization already impacted every part of our lives, but we are not at the end of the road. Namely, digital technologies are developing very rapidly. Education represents a significant part of human and especially of childrens' life. Children are our future, and they will create the future. Therefore, the education should be available, accessible, acceptable and adaptable. It should represent common good. The acquisition of digital skills needs should start from the earliest age (kindergarten) and should continue throughout our lives. Learning about digital technologies, digitalization, cybersecurity, robotics and artificial intelligence will be the backbone of future educational systems from primary level to life-long learning programs.

Because of the COVID-19 outbreak, children have been forced to continue the education process behind the screens and use digital technologies for learning. Their knowledge and autonomy have increased. But it is still necessary to proceed with learning about digital literacy. The inclusion on teaching digital literacy for children should be included more than ever before in educational curricula. It should be promoted in learning from an early age, especially in vulnerable groups such as Roma and children with disabilities. But to reach the digital transformation in

[44] Sun (2020), pp. 75–93.

[45] Ibid.

education, it is necessary to ensure that also teachers will join themselves in using digital skills. They also need to be appropriately assisted and trained (Committee on Culture and Education 2018). Only teachers who have appropriate digital skills and digital literacy can significantly raise our children's digital literacy.

On the other hand, to assure digital literacy and reduce the digital divide, it is necessary to provide also access to digital technology, especially to the internet. This goal is planned for 2030. By 2030, universal, safe, inclusive and affordable internet access should be provided, which should achieve universal connectivity (United Nations – Secretary General 2020).

References

Bucataru AM (2016) Using the convention on the rights of the child to project the rights of transgender children and adolescents: the context of education and transition. QMHRR 3(1): 59–81

Committee on Culture and Education (2018) Explanatory statement - report on education in the digital era: challenges, opportunities and lessons for EU policy design (2018/2090(INI)). https://www.europarl.europa.eu/doceo/document/A-8-2018-0400_EN.html#title2.

Committee on Economic, Social and Cultural Rights (1999) General comment no. 13: the right to education (article 13). https://www.ohchr.org/EN/Issues/Education/Training/Compilation/Pages/d)GeneralCommentNo13Therighttoeducation(article13)(1999).aspx

Committee on the Rights of the Child (2021) General comment no. 25 (2021) on children's rights in relation to the digital environment. https://www.right-to-education.org/sites/right-to-education.org/files/resource-attachments/UN_CRC_General%20comment%20No.%2025%20%282021%29%20on%20children%E2%80%99s%20rights%20in%20relation%20to%20the%20digital%20environment_En.pdf.

Committee on the Rights of the Children (2013) General comment no. 14 (2013) on the right of the child to have his or her best interests taken as a primary consideration (art. 3, para. 1). https://www2.ohchr.org/English/bodies/crc/docs/GC/CRC_C_GC_14_ENG.pdf

Digital Freedom Fund (2020) Digital rights are human rights. https://digitalfreedomfund.org/digital-rights-are-human-rights/article-26-the-right-to-education/

DiMaggio P, Hargittai E (2001) From the 'Digital Divide' to 'Digital Inequality': studying internet use as penetration increases. Working paper #15, summer 2001 - Center for Arts and Cultural Policy Studies, Princeton University. https://web.archive.org/web/20171018171421/http://www.princeton.edu/~artspol/workpap/WP15%20-%20DiMaggio%2BHargittai.pdf

Dwawan S (2020) Online learning: a panacea in the time of COVID-19 crisis. J Educ Technol Syst 49(1):5–22

European Commission (2020) Communication from the Commission to the European Parliament, the Council, the European Economic and Social Committee and the Committee of the Regions: Digital Education Action Plan 2021-2027 - Resetting education and training for the digital age, Brussels, 30.9.2020 COM(2020) 624 final. https://eur-lex.europa.eu/legal-content/EN/TXT/?uri=CELEX:52020DC0624

European Commission (2021a) Communication from the Commission to the European Parliament, the Council, the European Economic and Social Committee and the Committee of the Regions - EU strategy on the rights of the child. Brussels, 24.3.2021 COM(2021) 142 final. https://eur-lex.europa.eu/legal-content/en/TXT/?uri=CELEX%3A52021DC0142

European Commission (2021b) Digital Education Action Plan (2021-2027) - resetting education and training for the digital age. https://ec.europa.eu/education/education-in-the-eu/digital-education-action-plan_en (18 July 2021)

European Parliament (2018) European Parliament Resolution of 12 June 2018 on modernisation of education in the EU (2017/2224(INI))
European Union Agency for Law Enforcement Cooperation (EUROPOL) (2020) Exploiting isolation: offenders and victims of online child sexual abuse during the covid-19 pandemic. https://www.europol.europa.eu/publications-events/publications/exploiting-isolation-offenders-and-victims-of-online-child-sexual-abuse-during-covid-19-pandemic
FAO (Food and Agriculture Organization of the United Nations) (2020) Sustainable development goals - millennium development goal 2: achieve universal primary education. http://www.fao.org/sustainable-development-goals/mdg/goal-2/en/
FAO (Food and Agriculture Organization of the United Nations) (n.d.) FAO and the eight-millennium development goals. http://www.fao.org/3/a-az950e.pdf
Giannini S, Jenkins R, Saavedra J (2021). Mission: Recovering Education 2021. https://gemreportunesco.wordpress.com/2021/03/30/mission-recovering-education-2021/
Hajrullaia H, Saliua B (2016) The application of 4-A scheme in the context of higher education in Macedonia. Proc Soc Behav Sci 232:70–74
Human Rights Council (2020) Right to education: impact of the COVID-19 crisis on the right to education; concerns, challenges and opportunities. A/HRC/44/39
Human Rights Watch (2016) The education deficit failures to protect and fulfill the right to education in global development agendas. https://www.hrw.org/sites/default/files/accessible_document/educationdeficit0616_accessible.pdf
Human Rights Watch (2021) Years don't wait for them - increased inequalities in children's right to education due to the covid-19 pandemic. https://www.hrw.org/report/2021/05/17/years-dont-wait-them/increased-inequalities-childrens-right-education-due-covid
Institute for Higher Education in Latin America and the Caribbean (IESALC) (2020) COVID-19 and higher education: today and tomorrow; Impact analysis, policy responses and recommendations. April 9, 2020. https://unesdoc.unesco.org/ark:/48223/pf0000375693
International Covenant on Economic, Social and Cultural Rights, adopted and opened for signature, ratification and accession by General Assembly resolution 2200A (XXI) of 16 December 1966 entry into force 3 January 1976, in accordance with article 27. https://www.ohchr.org/en/professionalinterest/pages/cescr.aspx
Kaur A, Singh K (2014) Role of teacher education in the achievement of MDGs. Int J Eval Res Educ 3(2):125–132
Klees SJ, Thapliyal N (2007) The right to education: the work of Katarina Tomasevski. Comp Educ Rev 51(4):497–510
Kraljić S (2020) Implementation and protection of the child's right to education. Šolsko polje 31(3/4):27–44
Lopez-Guzman M (2020) Newton's "Year of Wonders" during the great plague. https://www.nysoclib.org/blog/newton%E2%80%99s-%E2%80%9Cyear-wonders%E2%80%9D-during-great-plague
Monteiro AR (2010) The right of the child to education: what right to what education? Proc Soc Behav Sci 9:1988–1992
Odink I (2019) Children's rights in the EU - marking 30 years of the UN Convention on the Rights of the Child. European Union - European Parliamentary Research Service. https://www.europarl.europa.eu/RegData/etudes/BRIE/2019/644175/EPRS_BRI(2019)644175_EN.pdf
Recommendation CM/Rec(2018)7 of the Committee of Ministers to member States on Guidelines to respect, protect and fulfil the rights of the child in the digital environment. https://search.coe.int/cm/Pages/result_details.aspx?ObjectID=09000016808b79f7
Sanders KC, Scanlon E (2021) The digital divide ss a human rights issue: advancing social inclusion through social work advocacy. J Hum Rights Soc Work 6:130–143
Schwab K (2017) The fourth industrial revolution. World Economic Forum, Geneva
Šimenc M, Kodelja Z (2015) Vsebina človekove pravice do izobraževanja. Šolsko polje 26(3–4): 195–215

Skulmowski A, Rey GD (2020) COVID-19 as an accelerator for digitalization at a German university: establishing hybrid campuses in times of crisis. Hum Behav Emerg Technol 2: 212–216. https://doi.org/10.1002/hbe2.201

Smith CW (2021) Consequences of school closure on access to education: lessons from the 2013–2016 Ebola pandemic. Int Rev Educ 67:53–78

Sun H (2020) Bridging the digital chasm through the fundamental right to technology. Georgetown J Poverty Law Policy XXVIII(1):75–93

Tawil S, Cougoureux M (2013) Revisiting learning: the treasure within assessing the influence of the 1996 Delors Report. UNESCO, Paris

Tomaševski K (2009) Right to education primers no. 3: human rights obligations: making education available, accessible, acceptable and adaptable. Novum Grafiska AB, Gothenburg

Tsolou O, Babalis T, Tsoli K (2021) The impact of COVID-19 pandemic on education: social exclusion and dropping out of school. Creat Educ 12:529–544

UNDP Bangladesh (n.d.) A situation analysis report on Education (MDG 2) – Bangladesh - a baseline for needs assessment and costing. https://planipolis.iiep.unesco.org/sites/planipolis/files/ressources/bangladesh_mdg_situation_analysis.pdf

UNESCO (2015) Rethinking education - towards a global common good? UNESCO, France

UNESCO (2020a) COVID-19 impact on education. https://en.unesco.org/covid19/educationresponse

UNESCO (2020b) Starting digital divides in distance learning emerge. https://en.unesco.org/news/startling-digital-divides-distance-learning-emerge

UNESCO (2020c) Education in a post-COVID world: nine ideas for public action International Commission on the Futures of Education June 22, 2022. https://en.unesco.org/news/education-post-covid-world-nine-ideas-public-action

UNESCO – Institute for Statistics (2021) Global monitoring of the school closures caused by COVID-19. http://covid19.uis.unesco.org/global-monitoring-school-closures-covid19/country-dashboard/

UNICEF (2019) Digital literacy for children: exploring definitions and frameworks. https://www.unicef.org/globalinsight/media/1271/file/%20UNICEF-Global-Insight-digital-literacy-scoping-paper-2020.pdf

United Nations – Generally Assembly (2016) Report of the Special Rapporteur on the right to education. A/HRC/32/37 https://undocs.org/A/HRC/32/37

United Nations – Secretary General (2020) Digital divide 'a Matter of Life and Death' amid COVID-19 crisis, Secretary-General Warns Virtual Meeting, Stressing Universal Connectivity Key for Health, Development SG/SM/20118 (11 June 2020). https://www.un.org/press/en/2020/sgsm20118.doc.htm

Universal Declaration of Human Rights, proclaimed by the United Nations General Assembly in Paris on 10 December 1948 (General Assembly resolution 217 A). https://www.un.org/en/about-us/universal-declaration-of-human-rights

Vuorikari R, Velicu A, Chaudron S, Cachia R, Di Gioia R (2020) How families handled emergency remote schooling during the Covid-19 lockdown in spring 2020. Publications Office of the European Union, Luxembourg

World Bank Group (2020) The COVID-19 pandemic: shocks to education and policy responses. https://openknowledge.worldbank.org/bitstream/handle/10986/33696/148198.pdf?sequence=4&isAllowed=y

Zhang D, Li X, Xue J (2015) Education inequality between rural and urban areas of the People's Republic of China, migrants' children education, and some implications. Asian Dev Rev 32(1): 196–224

Part V
Challenges in Achieving SDG 16

Corporations and the Global Backlash Over Privacy: Too Big to Regulate?

Eduardo Saad-Diniz and Matheus Puppe

Abstract Privacy design and (global) regulation of new technologies is an emerging concern for businesses, governments, and especially for the legal field itself. This article tries to explore strategies for controlling communication and information flows on how to regulate them and avoid abuses and the breach of our basic fundamental rights, as such privacy. Even more, when we refer to Artificial Intelligences and how it can escape our control (and its legal capacity to adapt its own normativity). Especially when considering cross-border legal issues, technology is the desired factor, and perhaps the corporations leading such advances are becoming too big to be regulated. Thus, digital compliance emerges as a relevant tool for an autopoietic process of self-regulation, in addition to the creation and even enforcement of this new type of law, triggered by the advances of social responsibility: social media and the respective reputational damage arising from "digital cancellations" and other freshly new sanctions.

Keywords Data Privacy · Compliance · Digital Law · Matheus Puppe · Privacy · LGOD · GDPR · Corporations · Technology

1 Introduction

The latest technological innovations are forging completely new political, moral and legal question. Those are holding such a disruptive potential that may challenge the own legal system's differentiation capacity. New advances, driven by an interconnected and globalized society, are developing faster than ever. Henceforth, traditional legal though are no longer providing the proper "normativity" to frame

E. Saad-Diniz
University of São Paulo (USP), São Paulo-SP, Brazil

M. Puppe (✉)
Goethe-University of Frankfurt, Frankfurt am Main, Germany

L. V. Sannikova (ed.), *Digital Technologies and Distributed Registries for Sustainable Development*, Law, Governance and Technology Series 64,
https://doi.org/10.1007/978-3-031-51067-0_9

the upcoming changes.[1] *Privacy design* and (global) regulation of new technologies is an emerging concern for businesses, governments, and especially for the legal field itself.

Major global players are contributing to the global economic development, but these tech giants are growing independently from a state border. The non-physical characteristics of those companies are enabling them to perform a role that shops through several jurisdictions, regardless of the fact that the domestic regulatory policies can no longer keep up with the disruptive communications of modern corporations and upcoming new technologies. The problem consists precisely in the fact that they operate in a "almost regulation-free zone", since cutting-edge innovations challenges the own ability to predict and secure operations. Perhaps, under a fine tune, those problems can be exacerbated by the extraordinary flow of information, not just between companies but also between users, raising even biggger questions about privacy and information use.[2]

In the regulation of communication (macro) and information flows (micro), digital compliance programs are getting specially relevant. Based on those programs management system improvement becomes possible, implementing mechanisms of detection, investigation and reaction to privacy violations, increasing the level of information security in the business environment or within the public sector. From a digital compliance programs perspective, it is possible to establish the boundaries between, on the one hand, authentic corporate behavior and voluntary collaboration between the company and public authorities, and, on the other, the violation of information security that pierce our legal rights, with emphasis in privacy. Ultimately, this "control" of communication and information flows are what really matters, as in the end, by the time of legal action before a court (*Verrechtlichung*)[3] the construction of truth within the justice system is based on those.

Of course, there is still a lot to be done in this field. Firstly, because digital transformation of social control studies are still neglected. The volume of information—and even the automation in the combination and generation of data through artificial intelligence—makes the traditional reading of the exercise of social control anachronistic. In some cases (see below on US vs. Europe Privacy Shield), regulatory policies or even the use of extensive self-regulation and compliance resources are already insufficient to control abuses, rights violations or even data leaks.

From a more realistic perspective, there is a process of "normalization" of this expropriation of data, in such a way that the social control of communication becomes the "control of communication that has already been violated". The extraordinary amount of data produced, does means that the control should be re-thought as a "control to prevent further right's violations (already violated)", especially in the verification of fundamental rights in business communications.

[1] Teubner (1997).

[2] Pagallo (2017).

[3] Teubner (1997). Ibid.

Therefore, this a field that must advance because the very understanding of compliance programs are up to now insufficient. When looking at the Brazilian compliance experience, there are very few evidences on the effectiveness of the use of private resources, as a way to express authentic ethical business behavior, or even on the need to implement certain controls.[4] There is no possibility of coherence in a regulatory policy and enforcement strategies still centered on individual accountability, without greater attention to the domestic order, in deep need to improve enforcement mechanisms in relation to the companies themselves.[5] Compliance initiatives, more often than ever, meet industry standards of compliance, with little influence on decision-making processes,[6] or are used as an overcompliance strategy, imposing market dominance through increased transaction costs.[7] Meanwhile, corporate concerns are justified by FCPA intimidation, not even in North American discussions there are greater clearness about the need, effectiveness and legitimacy of compliance programs.[8]

More specifically in digital compliance, a certain irrationality persists in the practices and discourses of the industry, there are almost no convincing empirical demonstrations about effective mechanisms for protecting privacy. Even so, regardless of the legal issues arising from this matter, new technologies and companies are still contributing to socioeconomic development. Technology helps us to improve accountability, brings information and education, allows better time management, increases productivity, longevity and comfort in life like nothing ever before. Given an expanding population, these advances are essential for human development (and even for our survival).

In light of this context, this essay tries to explore these strategies for controlling communication and information flows on how to regulate them and avoid abuses and the breach of our basic fundamental rights, as such privacy. Even more, when we refer to Artificial Intelligences and how they can escape to our control (and its legal capacity to adapt its own normativity). Especially when considering cross-border legal issues, technology is a desired factor, and perhaps the corporations leading such advances are becoming too big to be regulated. Thus, digital compliance emerges as a relevant tool for an autopoietic process of self-regulation, in addition to the creation and even enforcement of this new type of law, triggered by the advances of social responsibility: social media and the respective reputational damage arising from "digital cancellations" and other freshly new sanctions.

[4]On the authenticity and effectiveness of corporate behavior and its links to formal social control of business. Laufer (2006).

[5]de Susana Aires (2021).

[6]Saad-Diniz and Urban (2021), pp. 41–60.

[7]Rorie (2015), pp. 65–101.

[8]Lacerda (2021).

2 Privacy Shield Lessons

Privacy Shield was a framework created to regulate the transatlantic exchanges of personal data for commercial purposes between the European Union and the United States, with the aim of allowing American companies to properly receive personal data from EU entities in accordance with the European GDPR. The EU-US Privacy Shield replaced the International Privacy Principles, known as the Safe Harbor, which was declared invalid by the European Court of Justice in October 2015. However, in July 16, 2020, the CJEU declared the EU-US Privacy Shield invalid.

Recently CJEU, the highest court in the EU, issued a decision in the *Data Protection Commissioner v. Facebook Ireland and Maximillian Schrems* case, voiding the Privacy Shield as an acceptable method of exporting data for EU companies to the US, but retaining standard contractual clauses (SCCs) as an acceptable method—with one big caveat. Despite advances, through decisions such as these we observe that data protection aspects are still an emerging global discussion, in constant evolution, and still uncertain as per the (frequent) changes in new technologies and as a result, changes in parameters and decisions.[9] For thousands of companies that have gone through the *EU-US Privacy Shield* self-certification program, it means that they will need to review their framework, and it will certainly not be the first or the last time a situation like this occurs.

Certainly, data protection permeates contractual aspects, and like several other new technologies, international contracts are now being used as the necessary legal tool to (self) regulate relations between private parties (or even between public parties). A new *lex mercatoria* emerges in this uncertain context for purely state regulations.[10]

Among the main global dilemmas we can find the regulatory choice between social control or market deregulation. Corporations evolve in a highly dynamic and exponentialy growing. As such, disruptive technologies (such as recent advances in Artificial Intelligence) are changing the normative structure of our societ(ies). Surely due to their very own nature and due to the "social contract", States are unable to evolve and keep up with the same pace, which prevents an effective and meaningful regulation of the new global developments.

Further on this path, compliance programs come into scene, not only as a way to prevent and manage legal and social risks, but as a new autopoietic process of "self-constitutionalization" and self-creation of global law, as a result of a brand new *lex mercatoria*. Of course, this does not mean a corporative *laissez-faire*, but, at least, a proposal for self-regulation subjected to a regulatory policy articulated with intensive verification of fundamental rights compliance. More over, from the verification of the violation of privacy, such as the new regulation of the use of Artificial

[9]Case C-311/18, Data Protection Commissioner v. Facebook Ireland Limited and Maximillian Schrems, ECLI:EU:C:2020:559 (July 16, 2020), http://curia.europa.eu/juris/liste.jsf?num=C-311/18.

[10]Koskenniemi (2004).

Intelligence by the European Union, with several points that touch the core principles of data protection regulations.

In addition to the formal governance structures for data processing, there is no greater clarity regarding the use of compliance resources to protect privacy. Firstly, because neither corporative initiatives nor the national order offers a safer normative reference in which the protection of personal data differs from voluntary communication. The protection of privacy must not be confused with the judgment of opportunity and legitimacy of collaboration with public authorities. Apparently, the LGPD incurs in the same vices of imprecise notions, such as public interest or protection of the private against "invasion by third parties", demanding even greater precision with regard to the verification of fundamental rights by the corporations and in the business environment itself.

3 GDPR Vs. LGPD: Brasil-EU Connections

The European GDPR recognizes integrity and confidentiality as fundamental principles of data protection, reaffirming that personal data must be processed in a way that ensures adequate security of personal data, including protection against unauthorized or illegal processing, and loss, destruction or damage, using the appropriate technical or organizational measures even prior to the technical conditions. Moreover, the Brazilian LGPD recognizes security as a fundamental principle of data protection by affirming that this "security" means the use of technical and administrative measures capable of protecting personal data against any unauthorized access.

In Brazil, however, from the LGPD and the National Data Protection Agency (ANPD) created by it, very little can be said about more substantial changes in the exercise of data processing control. Perceptions surveys, involving IT professionals, show that a significant portion of business practices are still driven by notions of good faith or by "principiological" guidance from the LGPD. Organizations have not yet had the opportunity to develop more consistent techniques for handling personal data, internal communication processes and crisis management strategies (with proper preparation to respond to potential breaches, or leaks) are largely unknown. Little or nothing is empirically demonstrated to allow the identification of a new parameter or revision of the cooperative posture centered on the protection of privacy. In the absence of more concrete mechanisms, "privacy" is added to the conventional "public interest" already quite common in the compliance industry.[11]

The cooperation between legal and natural persons with public authorities is very uncertain, holding much more divergences than the actual act of sharing resources and control strategies.[12] It is not very different with the LGPD, as the sanctioning

[11] Saad-Diniz (2021) and Puppe (2021).

[12] Laufer (2018), pp. 67–802.

system reproduces a structure and functions quite similar to the domestic anti-corruption model, which are still vague in respect to a cooperative behavior with public authorities.

In any case, both laws are based on the need for prior consent, guaranteeing that data holder has the control over their own information. To do so, they consider some fundamental principles in data processing. Often, when it is said that national legislation has sought to align itself with European legislation, almost nothing makes reference to the fact that the GDPR has very unsatisfactory practical results. Processing of personal data is restricted to technical debate about the collection, storage and use of personal data. However, much more delicate than that, it faces a series of obstacles. On one hand, increased bureaucracy, which therefore leads to increased transaction costs. On the other hand, a high volume of information and automated decisions have shown that data processing depends on the capacity for self-control and, in most cases, does not overcome the cognitive bias.

This difficulty in establishing a regulatory policy, combined in a more intelligent way to exercise social control, ends up reflecting, in fact, that the transformation of the protection of fundamental our rights. It is the checks and balances to control de use and violations of the fundamental rights in order to prevent them to be further violated, orthe development of control mechanisms to ensure that"less" rights are violated.

Schematically, this principlism even seeks a certain correspondence with the constitutional text. In the Brazilian LGPD, there are ten principles:

1. *Purpose of the data*;
2. *Purpouse adequacy*;
3. *Need* for the treatment, hence collecting only the data necessary to be achieved based on the desired purpose;
4. *Free access* to your own data and to the mechanism on how it is treated;
5. *Data Quality*, keeping it accurate and up to date;
6. *Transparency*, especially regarding to the owner of the data;
7. *Safety*, adopting the minimum adequate technical measures;
8. *Prevention* against any misuse and damage to the data;
9. *Non-Discrimination*, preventing abusive or illegal use of the data;
10. *Accountability* of companies, forcing them to take the necessary measures.

The European GDPR, on the other hand, establishes seven principles determined in its article 5(1) and an additional one determined by article 5 (2):

1. *Legality*, being the data legally processed;
2. *Fairness*, data being processed fairly;
3. *Transparency*, with data being processed transparently;
4. *Limitation of Purpose*, or data collected for specific, explicit and legitimate purposes and not further processed in a manner incompatible with those purposes;
5. *Data Minimization*, or the collection of only adequate, relevant and purpose-limited data;
6. *Accuracy*, or data always exact and, when necessary, updated;

7. *Integrity and Confidentiality*, with data being processed in a way that ensures adequate security of personal data;
8. *Responsibility*, where the controller must be held responsible for observing the above seven principles.

Although distinct at a first glance, those principles are very much alike on their is essence, being a basis for compliance with both laws, and parameters for new developments, such as future regulations—as such the Artificial Intelligence regulation. More substantial measures to control privacy, protecting it against invasion by third parties, should be further explored, providing more details for still unclear situations, such as the consent of the data owner, information export, or even protection against leaks. The LGPD offers the possibility of structuring a privacy governance program, targeting information security (art. 50) and measures dedicated to the processing of personal data and protection of the interests of third parties.

For now, there are more difficulties than market practices that present better yields. The emphasis on consenting techniques in the use of information presents issues not only at the micro level—as such the difficulties of managing and controlling information—but also replicated in scale at a macro level, when facing multiple jurisdictions in which the communication of this information can be seen. The appointment of a data protection officer is more than debatable, especially if, instead of its real need, it does not only serve as a unnecessary transaction costs increase that do not affect ethical business behavior, not to mention persistence, in our order to, around the overload of normative expectations allocated to certain positions in the company hierarchy.[13] Corporations, at least in theory, should take those principles as a reference: digital compliance—corporate defenses and intensive verification of fundamental rights.

The LGPD's interpretation is quite dubious not only for its merely indicative character, but also for the same derivation of the compliance industry and motivated by the logic of reputational damage, business and market losses. Precisely—now regulated—compliance with the basic principles of privacy within this new fundamental right, encourages the use of new tools and technologies for the necessary technical adjustments.

We might observe, due to the same context of digital transformation, that the the evolution of new technologies designed and architected for the control and security of information, such as the (not so recent, but widely used) blockchain. A blockchain is a system that allows us to track the "sending and receiving" of some types of information over the internet, in a essentially incorruptible way. They are interconnected blocks of data or code fragments generated online containing information created as a "chain" (hence the name "block chain"). The technology was initially created to be a peer-to-peer electronic financial system, like a network that times transactions by placing them in a continuous chain in the hash, forming a record that cannot be changed without redoing all the work. Thus, for a better

[13] Saad-Diniz (2019).

understanding, we can think that, *because* they are connected, if someone unravels one code, they would need to unravel the hole code of the all the other blocks, and so on, making the task almost impossible.

Such technology, although created for a cryptocurrency, opens a new frontier for privacy, creating technical mechanisms that might allow the maintenance of privacy and security in an environment where this has been judged as mostly impossible. With this, data protection starts to gain a new technical viability to ensure the effectiveness of the compliance program. Surely an important global and state-independent right.

4 Corporate Defenses and Intensive Verification of Fundamental Rights

In fact, we are moving towards a more principled state function. Since the existing mechanisms and the speed of the new technologies do not allows a proactive and preventive public action, but rather a reactive one, we rely now more than even on a fundamental principles and rights basis—as seen in the new fundamental right to privacy—, which acts as parameter for compliance programs or other self-regulation strategies.

A concrete example of this new process is the right to privacy. As in Brazilian Federal Constitution art. 5, item X, the right to privacy—part of personality rights and fundamental rights—, or in Art. 2 of the German Constitution—*Jedermann-Grundrechte*—, within personality rights (*Persönlichkeitsrecht*), the right to a private sphere and consequent right of self-determination of personal information (*Recht auf informationelle Selbstbestimmung*), or in art. 8 of the European Union, the protection of personal data. They are part of successive fission and self-regeneration movements that forged a new fundamental right: the right to privacy. In this way, the process would be equivalent to:

(i) First came the decisions of large tech corporations to indiscriminately use personal information, violating users' privacy. Media scandals such as those at Cambridge Analitica are unleashed, putting such decisions in check, now subject to public scrutiny;

(ii) Subsequently, the social (or market) sanction emerges, materialized in criticisms and "boycotts" towards those companies, directly impacting on their reputation - and obviously on profits;

(iii) Therefore, as an attempt to repair (or rather, to recover), corporations recognize such "errors", standardizing them in their internal guidelines, described within their codes of conduct, (in a self-generation process of a standard and new regulations). Subsequently, we obser the self-application by compliance, as a way to optimize and improve Social Accountability. After all, they ultimately depend on the *Users*.

(iv) Finally, the State begins to exercise its power and function, regulating such self-regulations, generating new laws such as GDPR in Europe and LGPD in Brazil, as a top-to-bottom process. At this point, a Fundamental Right is created and from it, new regulations and principles will be elaborated, such as the Artificial Intelligence regulations.

5 Conclusion

Corporations are growing and becoming global conglomerates. Neverthelss they are still relying on their *Users*. The interconnectivity has enabled cross-border actions, but it has also enabled the alignment of expectations and social concerns globally, forging a new type of cosmopolitan citizen, as proposed by Kant, but now driven by and through social media. Too big to regulate? Probably, if we consider the traditional state perceptions and regulations, but social accountability starts to gain strength, and as such a "ant effort", society, now global and interconnected, through the new perceptions, are imposing new forms of regulation upon companies. Those companies, on the other hand are "capturing" such aspirations through their global codes of conduct, which are systematized within their compliance processes, gaining strength and importance within this context of parastatal business laws. The new principles are now guiding this process in which States, Corporations and Societies work together to understand and parameterize new and unpredictable situations arising from the most advances of technologies. And, in any case, trends in digital compliance do not outweigh the need for scientific assessment of the social control of corporative communications, especially in regard to the quality and consistency of the information as an expression of ethical corporative behavior.

References

Case C-311/18, Data Protection Commissioner v. Facebook Ireland Limited and Maximillian Schrems, ECLI:EU:C:2020:559 (July 16, 2020), http://curia.europa.eu/juris/liste.jsf?num=C-311/18

de Susana Aires S (2021) As diferentes faces dos programas de *compliance*. In: Martín AN, Saad-Diniz E (eds) Legitimidade e efetividade dos programas de compliance. Tirant, São Paulo

Koskenniemi M (2004) "Global Governance and Public International Law" https://www.nomos-elibrary.de/10.5771/0023-4834-2004-3-241.pdf

Lacerda NM (2021) Democratização da gestão empresarial segundo a perspectiva dos stakeholders e análise de risco: um caminho para efetividade dos programas de compliance? In: Martín AN, Saad-Diniz E (eds) Legitimidade e efetividade dos programas de compliance. Tirant, São Paulo

Laufer W (2006) Corporate bodies and guilty minds. Chicago Press, Chicago

Laufer W (2018) The compliance game. Revista dos Tribunais 988:67–802

Pagallo U (2017) Lecture at the specialization course Law and Technology, at Stockholm University

Puppe M (2021) Risk Management e compliance: o papel da gestão informação no controle dos riscos. In: Martín AN, Saad-Diniz E (eds) Legitimidade e efetividade dos programas de compliance. Tirant, São Paulo

Rorie M (2015) An integrated theory of corporate environmental compliance and overcompliance. Crime Law Soc Chang 64(2):65–101

Saad-Diniz E (2019) Ética negocial e compliance: entre a educação executiva e a interpretação judicial. Thomson Reuters Brasil, São Paulo

Saad-Diniz E (2021) Falsa percepção do controle nos programas de compliance: pluralismo moral e análise de risco. In: Martín AN, Saad-Diniz E (eds) Legitimidade e efetividade dos programas de compliance. Tirant, São Paulo

Saad-Diniz E, Urban M (2021) Análise qualitativa dos programas de *compliance* no Brasil (2014-2019). Revista dos Tribunais 5:41–60

Teubner G (1997) Global law without a state. Dartmouth, Aldershot

E-Justice in Russia

Vasily Laptev

Abstract Electronic justice in Russia is part of the general trend of digitalization of public authorities. The present work reveals the main elements of electronic justice in Russia at the current stage. The paper examines trends and patterns in the development of electronic justice focusing on the needs of civil society and the business community. It also explores positive and controversial aspects of the introduction of digital technology in court. The study provides an overview of certain aspects of digital justice, requiring the state to ensure equal technological accessibility and elimination of digital inequality among participants of the judicial process. The research formulates the limits of the use of information technology in court, considering the technological accessibility to the electronic court. The present study forms a holistic view of the state of Russian electronic justice.

Keywords Electronic justice · Electronic document management · Electronic evidence · Online courts · Remote hearings · Court bot · Court cloud · Artificial intelligence · Cyborg judge

1 Introduction

The development of electronic justice, as a priority in the activities of Russian courts, was noted at the IX All-Russian Congress of Judges (2016). Within the framework of this concept at the initiative of the President of the Russian Federation and the Supreme Court of the Russian Federation a number of laws were adopted. As a result of the reform of procedural legislation, the efficiency of justice has increased.

The article was previously published in the journal Law & Digital Technologies, 2022, 2(1).

V. Laptev (✉)
Kutafin Moscow State Law University (MSAL), Moscow, Russia

L. V. Sannikova (ed.), *Digital Technologies and Distributed Registries for Sustainable Development*, Law, Governance and Technology Series 64,
https://doi.org/10.1007/978-3-031-51067-0_10

The Concept of Judicial Information Policy for 2020–2030, approved by the Council of Judges of the Russian Federation on December 5 2019, laid the foundation for improving the ways in which citizens, organizations, public associations, state and municipal bodies, and media representatives can access information about the activities of the courts. The Concept of Informatization of the Supreme Court of the Russian Federation dated February 2, 2021, No. 9-P (Supreme Court of the Russian Federation 2021)[1] defines directions for improving modern digital justice.

The issues of digitalization of Russian judicial proceedings have become the subject of numerous scientific and practical studies of legal scholars and practicing lawyers (Laptev and Solovyanenko 2017, 2019a).[2] The constant development and variability of technological equipment of courts in Russia is noted in the periodical and specialized press, as well as in interviews with leading legal experts.

Digital justice has significantly facilitated the workload of the Russian courts. It made it possible to structure information on court cases, keep quick statistics and respond to issues arising in the work of the courts.

Powerful digital servers and archives (operational and backup) ensured reliable electronic storage of documents related to the activities of the Russian courts.

2 General Provisions on Digital Justice: Concept and Contents

In Russia, the terms "electronic justice" and "digital justice" are often used as identical concepts. In practice, the terms "mobile justice", "remote justice", "cyberjustice", "online justice", and other definitions of traditional justice based on information technologies are also interchangeable.

Digital justice is a form of law enforcement activities of the judiciary to implement constitutional, civil, administrative and criminal proceedings using information and communication technologies and systems, including a single information space of courts, software for the automation of court proceedings (filing, processing and scanning of documents received by courts, formation of electronic cases and electronic archive of court cases, online sessions, artificial intelligence, and others), providing for the use of information and communication technologies.

Public speeches of the Chairman of the Supreme Court of the Russian Federation V. Lebedev testify to the accelerated pace of transition of domestic courts to the electronic system of justice in the context of technological and other present-day challenges (RAPSI 2015; Alexandrov 2020).[3]

[1] Supreme Court of the Russian Federation (n.d.).

[2] Laptev and Solovyanenko (2017), pp. 16–21; Laptev and Solovyanenko (2019a).

[3] RAPSI (2015) and Alexandrov (2020).

Several legal scholars, in particular Romanenkova (2013),[4] interpret e-justice (in the broad sense) as a set of various automated information systems—services that provide means for publishing judicial acts, conducting the "e-case" and the parties' access to the materials of the "e-case". Sharayev (2010)[5] believes that e-justice can be seen as a way of administering justice based on the use of information technologies. Quite often the category of "e-justice" is disclosed through the enumeration and description of its separate elements (reflecting the openness of justice, electronic interaction between the participants of the process and the electronic case work in court) (Sharifullin et al. 2018).[6] The above positions illustrate the basis of e-justice, for example, the digital technology.

Russian digital justice includes essential elements which were formulated and enshrined in the Concept of the Federal Target Program "Development of the Judicial System of Russia in 2013 - 2020" dated September 20, 2012, No. 1735-r; the Federal Target Program "Development of the Russian Judicial System for 2013-2020" dated December 27, 2012, No. 1406, the Concept of Judicial Information Policy for 2020–2030 dated December 5, 2019, and other acts. The increased number of cases heard in courts, which cannot be administered in traditional ways without the use of information technology, became a significant incentive for the development of e-justice. Thus, in 2020, courts of arbitration at all levels heard 1.5 million cases, and courts of general jurisdiction of first instance heard 29.5 million civil and administrative cases, as well as 530,000 criminal cases (Supreme Court of the Russian Federation n.d.).[7]

The components of digital justice are:

- court automation system[8];
- electronic document management, including electronic filing and processing of documents (statements of claim, complaints and other petitions) submitted to the court in electronic form;
- scanning of incoming documents;
- formation of electronic files and archives;
- automated analytical support of the work of the courts;
- personal accounts of those involved in the process;
- establishment of technical conditions for ensuring electronic interaction of the courts of general jurisdiction with the information systems of the Office of the Procurator-General of the Russian Federation, the Ministry of Internal Affairs, the Federal Bailiff Service, the Federal Penal Correction Service and others;

---

[4] Romanenkova (2013), pp. 26–31.

[5] Sharayev (2010).

[6] Sharifullin et al. (2018), pp. 57–62.

[7] Supreme Court of the Russian Federation (n.d.).

[8] Software and hardware complex providing for the conduct of court proceedings, recording in electronic form the results and progress of procedural actions performed by the court and other participants in court proceedings

– courts and the Judicial Department of the Supreme Court of the Russian Federation will be equipped with software and key media for electronic document management with the use of an electronic signature.

The following elements are expected to emerge soon:

- the creation of a cloud computing architecture to provide for the automation of court and general office workflow;
- expanding the use of mobile devices as access to information resources, software systems and databases of arbitration courts using "cloud" technology for judges and court officials—mobile justice;
- establishment of mobile judges' offices (special passenger minibuses) with the use of videoconferencing for mobile sessions in geographically remote areas of the country;
- organization of incoming scanning of all documents received by arbitral tribunals and the formation of electronic cases (pilot project "Electronic Case");
- creation of an electronic archive for subsequent transition to electronic enforcement of judicial acts;
- integration of court information platforms with information systems of public authorities;
- creation of specialised legal electronic libraries in the courts;
- creation of a single digital space of electronic signature trust is the basis for the future development of e-justice, including the single information space of the judicial system;
- recording of the course of proceedings by means of audio recordings and the inclusion in the case file of discs, flash cards and other evidences in digital form, in accordance with the principle of one electronic data carrier for each case separately, and their cloud storage;
- granting participants in the proceedings the right to consult case files in a cloud-based repository;
- development of electronic identification and other means to ensure the security and reliability of electronic court procedures.

The current situation shows that there is no possibility of a full transition to digital justice in Russia with all its elements. For a long time, there will be a mixed system of justice, which includes elements of the traditional paper-based and electronic document flow. This is due to the digital divide.

Arbitration courts considering economic disputes may soon completely switch to digital justice, as it is assumed that most participants in a dispute are professionals representing business interests, economic entities, government authorities and other parties, who usually possess minimal technical equipment for using electronic document management. Analysts estimate that Russia has a high level of informatization of economic justice, with indicators comparable with those of

Singapore and China and exceeding those of Australia, Germany and Canada (Kashanin 2020).[9]

For courts of general jurisdiction which hear disputes, including those involving citizens (pensioners, minors, people who have lost legal capacity or live in places which are not easily accessible), there is a lack of material and technical capacity for full electronic interaction with the court.

Digitalization of justice should take place in conditions of real possibility (accessibility) of information technologies for the participants of the process. A different approach violates the constitutional guarantee of judicial protection of rights and freedoms (Article 46 of the Russian Constitution).

3 Electronic Document in Court

The gradual transition from paper documents to digital documents in court proceedings has enhanced the ability of litigants and the judiciary to exchange digital information of legal significance. Key issues of electronic case management were clarified in the Resolution of the Plenum of the Supreme Court of the Russian Federation dated December 26, 2017 No. 57 "On certain issues of application of legislation governing the use of documents in electronic form in the activities of courts of general jurisdiction and arbitration courts".

Two types of documents in electronic form are distinguished in the judicial electronic document management system in Russia: an electronic image of a document and an electronic document. The electronic image of a document (an electronic copy of a document produced on paper medium) is a copy of a document produced on paper medium converted into electronic form by means of scanning, certified in accordance with the Procedure of document filing with a simple electronic signature or an enhanced qualified electronic signature. The electronic document is a document, created in electronic form without prior documentation on paper, signed with electronic signature in accordance with the legislation of the Russian Federation.

Depending on the legal consequences of the appeals all documents coming to the court can be divided into two groups. One group is documents that are required to be signed with an enhanced qualified electronic signature (electronic documents) or are certified their electronic images with an enhanced qualified electronic signature: an application for provisional security for the protection of copyright and/or related rights; an application for provisional legal protection; an application for provisional measures of protection.

The other group is documents that are not required to be signed with an enhanced qualified electronic signature (documents in electronic form signed with a simple electronic signature): an application; an administrative statement of claim; a complaint or a submission; documents attached thereto, and others. It should be

[9] Kashanin et al. (2020).

considered that documents not requiring an enhanced electronic signature can be sent to court both in the form of an electronic image of the document and as electronic documents signed with the enhanced qualified signature.

In practice, there are cases when electronic images of documents (produced on paper) that are simultaneously certified with an enhanced qualified electronic signature are sent to the court. Double authentication of a document (a simple and reinforced electronic signature) can be regarded as an additional measure to protect rights of the process participants and to ensure digital hygiene which is not superfluous in the context of cyber-security measures.

If it is not legally required to sign an application to the court with an enhanced electronic signature, then a power of attorney or other document, confirming the authority of the person to sign the application to the court in the form of an electronic image of the document certified by a simple electronic signature of the person submitting the documents to the court, shall be attached when filing an application to the court in the form of an electronic image of the document. Depending on the type of court proceedings, courts of different levels apply appropriate document management systems.

4 Electronic Workflow in the Court

Electronic filing of documents is governed by the relevant instructions and regulations. To file documents it is necessary to register a personal account, depending on the competent court:

- arbitration courts and the Supreme Court of the Russian Federation (within arbitration proceedings)—in the My Arbitrator system (https://my.arbitr.ru/#index);
- courts of general jurisdiction and justices of the peace—in the section "Submission of procedural documents in electronic form" of the GAS "Justice" internet portal (www.sudrf.ru);
- the Supreme Court of the Russian Federation—in the information system of the official website of the Supreme Court (http://supcourt.ru/);
- the Constitutional Court—in the "Appeal to the Constitutional Court of the Russian Federation" filing system (https://petition.ksrf.ru/).

As a special case, documents in electronic form can also be submitted to the Constitutional Court of Russia to the general e-mail address—ksrf@ksrf.ru with signing the application and its attachments with an enhanced qualified electronic signature, for example, in the form of an electronic document.

The personal account shall be registered in the name of the individual who submits the documents to the court electronically. If electronic submission of the document is done by the representative of an organisation, the personal account shall be registered in the name of the respective representative.

An electronic image (or the electronic copy) of a document should meet the following requirements: a paper document should be scanned at 1:1 scale in black and white or grey colour (with 200–300 dpi resolution); a PDF format is required (it is recommended that PDF text should be copyable); the size of an electronic image file should not exceed 30 Mb, and so on.

Electronic documents should satisfy the following requirements: must be initially created in an electronic form without preliminary documenting on paper-based media; PDF text must be copyable; the files of documents attached to the appeals to the court shall be presented in the format, in which they are signed with electronic signature; the file size of an electronic document shall not exceed 30 Mb; each separate document shall be presented as a separate file. The file name must allow for identification of the document and the number of sheets in the document.

Electronically filed documents shall be rejected by the court on the following grounds: the petition is not addressed to the court; the petition is identical to the previously filed petition; the documents are illegible, in particular: the pages of the document(s) are reversed; the document(s) do(es) not contain all pages; the presence of all pages is not identifiable (because of the absence of pagination); the file does not contain an electronic document or an electronic image of the document; incoherent text; missing file of the petition and/or files of the documents attached to it.

5 Digital Audio and Video Recording of Court Proceedings. Broadcasting the Session

Electronic video and audio court transcripts are essential in the Russian digital justice system. Depending on the accessibility of these digital transcripts for litigants and third parties (non-disputants), a distinction is made between: digital audio and video protocols of public access (for example, sessions of the Plenums of the Supreme Court of the Russian Federation); open-access digital audio records of an individual involved in a case; digital video records for official use (court sessions of arbitration courts of first and appellate instances).

Maintaining electronic audio and video records also secures the fundamental principles of court proceedings of publicity and openness.

The Russian procedural law provides that individuals present at an open court session have a right to take notes during the court session and audiorecord the session. Filming and photographing, videotaping, and broadcasting arbitration court sessions on radio, television and on the Internet is allowed with the permission of the judge presiding over the hearing.

Audio-recording of court hearings makes it possible to reflect on the participants' procedural actions and on the results of their examination by the court. The Arbitration Procedure Code of the Russian Federation does not provide for minutes to be kept during the preliminary hearing and the appeal proceedings. Nevertheless, all applications and motions of the individuals involved in the case and the results of their examination shall be contained in the case file and shall be reflected in the ruling or decision of the arbitral tribunal or in the record of the individual procedural action.

It should be taken into account that the digital audio-recording of the court session shall be regarded as the primary and basic means of recording the proceedings in court, despite the fact that a written record shall also be prepared as an additional means of recording.

Different courts use appropriate audio recording technology. Thus, in arbitration proceedings audio recording of each court session is carried out by connecting digital dictaphones and external microphones to the workplace; using audio-recording hardware and software, and others.

Videorecording of court hearings in arbitration courts allows to record a court session via videoconferencing. Videoconferencing improves the quality of work of the judiciary, as well as consideration of complaints about the actions (including ethics) of judges, and the like. As a rule, a video recording of the court session is not made available to the parties to the arbitration proceedings. The parties are entitled to receive copies of the recording of the videoconference of the court session. Within the framework of administrative proceedings, both audio and videorecording of court hearings may be made by virtue of Article 205 (4) of Code of Administrative Judicial Procedure of the Russian Federation. Accordingly, since these media obtained by technical means by the court shall be attached to the record, the said information may be provided as a digital copy at the party's request.

Broadcasting of hearings. The issues of publicity and openness of court proceedings are reflected in detail in the Federal Law "On providing access to information on the activities of courts in the Russian Federation" dated December 22, 2008, No. 262 by virtue of which citizens, including representatives of organisations, public associations, state and municipal bodies, have a right to attend open court proceedings and record the proceedings in the manner and form prescribed by Russian law (Article 12).

For Russian courts at all levels, the Procedure for organising and broadcasting court hearings via radio, television and the Internet, approved by Order of the Judicial Department of the Supreme Court of the Russian Federation No. 182 of October 17, 2017, is applicable. The decision on the admissibility of court session broadcasting shall be reflected in the protocol of the court session. Broadcasting may be time-limited by the court and shall take place at the court-appointed seats in the courtroom. Broadcasting a court session without authorisation shall be a breach of order in the court session.

According to Point 17 of the Resolution of the RF Supreme Court Plenary Session of June 15, 2010 No. 16 "On Practice of Application by the Courts of the RF Law 'On Mass Media'" the judge may not hinder the mass media from entering the court session and from covering a particular case, except for cases provided for by law (for instance, if the case is heard in a closed court session, if the media representatives have to leave the court room for violating the order in the court session and others).

6 Electronic Evidence

Electronic evidence is gradually gaining popularity in modern legal proceedings (Laptev 2017; Zhurkina 2020; Obidin 2020). Traditional evidence converted into a digital form (a postal letter converted into electronic communication, a bilateral written agreement and an electronic contract, photographs, and so on) possesses not only their negotiability and ease of presentation to the court, but also the collection and storage of said evidence including in the formation of court files and their archives.

The list of possible types of electronic evidence in court is open. With the development of online justice, the legislator must work diligently to put all sorts of technological tools in the hands of the judge in the administration of justice to establish the objectivity and authenticity of evidence. For example, deepfake (Brandon 2018; Sozankova 2019),[10] substitution of a domain name or an Internet portal, changing the content of a video clip, adjusting a web archive (a digital imprint of a website page) and other products of artificial intelligence may be a significant obstacle to the objective establishment of truth in a case by the court.

The issue of admissibility of a range of electronic evidence and the expediency of introducing the concept of "electronic evidence" into the law is of considerable import in the legal literature (Zhurkina 2020).[11] Issues of introducing information technology directly into criminal investigation, examination and adjudication procedures have also been investigated (Voronin 2019).[12] It has also been proposed to consider electronic information as an independent type (source) of evidence, as it includes not only information on facts but also digital requisites (for example, date of file creation, changes of its content, format and author) (Obidin 2020).[13]

6.1 Electronic Communications

Civil and commercial interactions are enabled by the exchange of information, through which the parties express their will to enter, change or terminate a legal relationship. Digital technologies have provided business entities and the legal communities with modern methods and means of electronic communication—electronic messages. To a large extent, electronic correspondence and electronic negotiations make it possible to determine the true will and intentions of the parties before the conclusion of the contract. Types of electronic communications are an e-mail, SMS messages, fax messages (Public Switched Telephone Network), voice mail

---

[10] Brandon (2018) and Sozankova (2019).

[11] Zhurkina (2020), pp. 44–46.

[12] Voronin (2019), pp. 74–84.

[13] Obidin (2020), pp. 198–206.

messages (Public Switched Telephone Network), Instant messaging (Instant messaging), SMTP messages and others.

Documents received by facsimile, electronic or other means of communication, including information and telecommunications network "Internet", as well as documents signed with an electronic signature in the manner prescribed by Russian law are admissible as written evidence.

6.2 *Electronic Documents and Other Documents in Electronic Form*

According to Russian procedural law, documents signed with an electronic signature (an electronic document or a scan copy of a paper document) are recognised (admissible) as written evidence. As of 2019, the new wording of Article 160 of the Civil Code of the Russian Federation includes the recognition of a handwritten signature in transactions of a facsimile reproduction of a signature by mechanical or other copying means or of another analogue of a handwritten signature in cases and in the manner provided by law, other legal acts or agreement of the parties.

By virtue of paragraph 2 of Art. 434 of the Civil Code of the Russian Federation, an agreement in writing may be concluded by a single document (including an electronic one), signed by the parties, or the exchange of letters, telegrams, electronic documents or other data. When a contract is concluded by means of an exchange of documents, for the purposes of recognising an offer as an offer, the offeror's signature is not required if the circumstances, in which the offer is made, allow reliable identification of the person who has sent it. Bilateral (multilateral) transactions may be carried out in the ways set out in paras. 2 and 3 Article 434 of the Civil Code of the Russian Federation, except for cases where an agreement has been concluded as a result of fraudulent actions.

6.3 *Digital Audio, Video and Photo Recordings*

Evidence on magnetic media demonstrating information on the facts—audio, video or photo images—has become an effective means of proof. This evidence is used in court along with traditional evidence such as witness statements, paper evidence, and the like. A clarification by the Supreme Court of the Russian Federation that for audio or video recordings to be considered as admissible evidence in cases on protection of infringed intellectual property rights the consent of an individual in respect of whom they are made is not required, but is necessary for participants in proceedings.

Evidence of digital audio and video recordings and photographic images is gradually being enshrined in legislation as a separate type of evidence. Thus, in

2016, Article 26.7 of the Code of Administrative Offences of the Russian Federation was supplemented with a provision which includes photo and film recordings, audio and video recordings, information databases and data banks and other media as "documents" as evidence of an administrative offence.

Evidence of legally relevant facts by means of video and audio recordings is also actively used in corporate practice when certifying decisions of meetings of participants in non-public corporations by virtue of Article 67.1 paragraph 3 of the Civil Code of the Russian Federation. Video and audio recordings of general meetings and meetings of the board of directors and other collegial corporate bodies ensure the implementation of the principles set out in the Corporate Governance Code of the Bank of Russia.

An interesting issue is the recognition of an audio recording of a court session in another dispute as "electronic evidence". It is known that statements of individuals involved in a case, including the statements given during a court hearing, are evidence. Furthermore, court consent for audio-recording of court hearings is not required.

Photographs, including those taken by digital devices, are sometimes recognised as the main and only evidence in cases in arbitration courts and courts of general jurisdiction.

6.4 *Information from Internet Portals of Public Authorities and Organisations (Printouts and Screenshots)*

Under current clarifications of the Supreme Court of the Russian Federation, admissible evidence includes, inter alia, printouts of materials posted on an information and telecommunications network (screenshot) made and certified by persons participating in the case, indicating the address of the website from which the printout was made, as well as the exact time of its receipt. Such printouts shall be evaluated by the court in the hearing of the case along with other evidence.

The following Internet portals shall serve as evidence: Registers of the Federal Tax Service of the Russian Federation (Unified State Register of Legal Entities; Unified State Register of Individual Entrepreneurs); Fedresource (portal Unified federal register of information on the activities of legal entities, individual entrepreneurs and other subjects of economic activity); Russian Post portal in the section for tracking the history of court correspondence by their tracking numbers; Unified State Register of Real Estate (Rosreestr); Business information resources.

6.5 SMS Messaging, Business Messengers and Social Networks

Recently, it has become customary for business entities' managers and employees to use SMS messages, as well as messages via business messengers and social networks (Skype, Viber, WhatsApp, Telegram, Signal, Facebook Messenger, Wire, Amo, Microsoft Teams, Twist, Discord, TamTam, Dialog messenger, Yandex Chats (ex. Yamb), Cisco Spark, Hangouts, Rocket Chat, Myteam, Zulipchat, VIPole Secure Messenger, Google MessagesICQ, Line and many others) as evidence in court. In particular, the courts recognize the fact of sending a pre-trial claim through messengers.

6.6 Electronic Copies (Image) of a Document

As a general rule, a copy of a document drawn up in a paper form may not serve as evidence of the relevant fact in the case in the absence of the original. The need to provide the original document initially produced in a paper form, as well as the right of the court to demand them, is reflected in paragraph 9 of Resolution No. 57 of the Plenum of the Supreme Court of the Russian Federation dated December 26, 2017. On the other hand, in some cases it may be obvious that the other party to the case (for example, former CEO in a corporate dispute, heirs in an inheritance dispute or spouses in divorce proceedings), having actual access to the original, has deliberately disposed of this evidence. Besides, it may happen that only an electronic copy is preserved, for example in PDF or JPEG format. In addition, the parties may initially agree to communicate electronically and to send electronic images (copies) of documents via e-mail and other means of telecommunication.

6.7 Digital Imprint of Website Content (Web Archives)

Digital technology stores the history of changes to the site content in DATA centres, making "digital imprints" of the relevant pages of the sites. Examples are foreign (http://web.archive.org/) and Russian sites (https://webarchiveorg.ru/, http://web-arhive.ru/).

7 Electronic Case Review (e-Review)

In the modern digital justice system, there are several ways of electronic familiarisation of the participants of the process with the case file: in cases considered under the simplified procedure; in cases considered under the general rules of action proceedings or proceedings on cases arising from administrative and other public legal relations. In the first case, the court must provide the parties and third parties in the case with a code of access to the electronic case file. It is not required to send the respective application to the court, and the access code is provided automatically by sending a postal order to accept the statement of claim and examine it in simplified proceedings. The second procedure of familiarisation involves an application to the court for familiarisation electronically with the case file.

Many arbitration courts in Russia have a procedure for examining case files electronically. The development of such a regime corresponds to the Instruction on record keeping in arbitration courts of Russia. The sanitary and epidemiological situation in Russia and around the world, as well as the measures taken to limit the spread of the new coronavirus infection, have been a major impetus for the development of this familiarisation regime. In addition, this procedure is in line with the concept of digitalisation of the domestic judicial system.

The procedure under consideration for electronic familiarization with cases considered under the general rules of action proceedings or proceedings in cases arising from administrative and other public legal relations operates in the Arbitration Court of the Moscow District. In the "My Arbitrator" service on the home page, the participant selects "Applications and petitions", enters the case number and on the "Type of application" tab selects "Application for familiarisation with case materials". Then, on the "Documents" tab of the application form, the case participant checks the "Grant access to the case file in restricted mode" box. Other fields of the form, which must also be filled in, are: registration of the petition by a court employee and automatic notification of the document registration by the system; approval of the petition by the judge hearing the case; granting 24 h access from the date of approval of the application, for which a notification containing the access code and the validity period of the code is sent to the participant in the case.

8 Court Noticed. Electronic Dispatch of Judgments

In Russia, different forms of court notices of the participants of the proceedings are used, depending on the type of court proceedings.

In arbitration proceedings, the information about accepting a statement of claim or application for proceedings, as well as the information about the time and place of the court hearing or a separate legal proceeding shall be posted by the arbitration court on their website no later than 15 days before the court hearing or a separate legal proceeding, unless otherwise provided by the procedural law.

In several arbitration courts, in particular, the Arbitration Court of Moscow, a service for sending judicial acts electronically in the form of registered letters through the Software Package "Litigation of arbitration proceedings" is available. Service allows to choose the necessary judicial act and send it by registered electronic mail, and defines its status ("sent", "returned", and so on).

9 Online Sessions. Internet Courts

Currently, there is an ongoing discussion on the choice of an efficient technological platform to provide online sessions in Russian courts. Historically, the remote mode of court hearings started to function through videoconferencing. However, court hearings via videoconferencing and online hearings are not identical.

Sessions using videoconferencing are provided by the specialists of the software and hardware department of the relevant courts (under the mandatory condition that the participants of the dispute visit the court building), while the online sessions are run through the information system "My Arbitrator", with the authorisation via Unified identification and authentication system, which allows to participate in the session without visiting the arbitration court building (from the office, home, and so on).

In order to prevent the spread of COVID-19, the Supreme Court of the Russian Federation recommended that courts in a number of categories of cases (in urgent situation), hold court sessions using a video-conferencing system and (or) a web-conferencing system, if it is technically possible, and if the opinion of participants of court proceedings have been taken into account.

In the case of a court session via web conference, participants of court proceedings submit to the court an application in electronic form with attachment of electronic images of documents proving their identity and authority. The Arbitration Court of the Yamal-Nenets Autonomous Area held its first online session on April 28, 2020.

Currently, the system of online sessions operates in 100 arbitration courts (out of total 118 arbitration courts), including the Supreme Court of the Russian Federation, all district arbitration courts and a number of courts of first and appellate instances. The Supreme Court of the Russian Federation has long practiced online sessions of the Plenum of the Supreme Court of the Russian Federation.

In Russia, online sessions are provided by the portal "My Arbitrator" through the "Online Meetings" tab.

10 Court Chat-Bot

A court bot (court robot) is a robotic program that allows to automatically answer the applicants' questions or perform other actions related to court proceedings.

In Russia, LegalTech-companies began to develop first legal chatbots (Zscheige 2020)[14] to help legal practitioners (for example, InfoCourt Bot, LawProInfoBot, Docubot and LawBot), which allow them to monitor the upcoming trials, change information on incoming claims, publish court acts on kad.arbitr portal, upload cases to calendar, create and analyze legal documents, interact with courts and state registries, and so on.

Specialised court chatbots are being developed. For example, the Pravo.ru team has developed a virtual assistant, the Arbitrator-bot. It is available on all e-justice service pages, and is trainable. The peculiarity of its work is the concreteness of the question. Thus, the more questions are asked, the more accurately the Arbitrator-bot answers.

11 The Judical Cloud

The issues of digital structuring and storing in the cloud services of information in electronic form are discussed in legal, economic and technical literature (Laptev and Solovianenko 2019b; Deshko et al. 2016; Reese 2009).[15] The development of "judicial cloud" seems promising for the domestic judicial system.

The technological structure of a cloud storage facility should have a reliable and secure architecture, including the appropriate level of Data Centres. Many existing cloud storage facilities on the Russian IT market (for example, Dropbox, Google Disk or iCloud) are administered in foreign countries by foreign corporations. Additionally, the actual storage (including a backup storage) of information in Data centres is also performed outside the Russian Federation. The need to develop Russian business in this area of information services has been reflected in the Strategy for the Development of the Information Technology Industry in the Russian Federation for 2014–2020 and for the period until 2025.

The following principles should guide the construction of a judicial cloud in Russia: technological access of Russian courts to the judicial cloud; guaranteeing the protection of digital information; backing up the judicial cloud (backup copy); location of the Data Centre(s) in Russia; a technological alternative in the management of the judicial cloud (Laptev and Solovianenko 2019b).[16]

[14] Zscheige (2020).

[15] Laptev and Solovyanenko (2019b), pp. 195–204; Deshko et al. (2016); Reese (2009).

[16] Laptev and Solovyanenko (2019b). Ibid.

12 Applications of Artificial Intelligence in Court

Two applications of artificial intelligence are suggested: court proceedings and general litigation issues; and estimation of evidence and determination of legally significant circumstances.

In revealing the spheres of AI application on general issues of court proceedings, the following factors are singled out: language of court proceedings, digital protocol, formation of court composition and category of cases, digital writs of execution, and so on.

Special questions, solved by AI, can be an assessment of evidence and establishment of legally significant circumstances in separate categories of disputes: legal assessment of the concluded deal, calculation of damages, determination of limitation periods, conciliation and amicable settlement, deepfake and other high-quality forgeries of evidence.

The development of digital technology in the era of information society and big data has proved the prospect of introducing artificial intelligence in court (Laptev 2021).[17] It has become apparent that artificial intelligence is our present, not the future as we have recently argued. However, there is still a long way to go for software engineers working together with neurobiologists to build an artificial cognitive system that approximates in structure and capabilities the human brain, which in turn has not been fully investigated by science.

The proposals of the leadership of the Supreme Court of the Russian Federation and the Council of Judges of Russia (Momotov 2020) regarding the gradual introduction of a "weak artificial intelligence" capable of solving highly specialised tasks in court are seen as timely.

Proceeding from the understanding of artificial intelligence as a simulated (artificially reproduced) intellectual activity of human thinking, the following stages of its introduction into the system of domestic courts can be suggested:

Short-term perspective: introduction of artificial intelligence as an assistant to a human judge on certain issues of case management and in the consideration of the merits of the case.

Medium-term perspective (5–10 years): allowing to consider artificial intelligence as the judge-human companion of the judge, including the question of estimation of several proofs.

Long-term perspective: possible replacement of the judge-human with an artificial intelligence on performance of separate functions of the judge-human at administration of justice.

The proposed forecast of stages of introduction of judicial artificial intelligence (judicial-AI) is based, first, on the level of information technologies development. Artificial intelligence, which approximates to cognitive abilities of human brain and billions of its neurons, has not been created yet. The computational function of

[17] Laptev (2021).

existing cyber-physical systems is effective, but still narrowly focused. It will take a long time to create an objective, erudite and "intelligent" forensic neural network.

It seems that soon, AI will recognize and translate into Russian (the language of court proceedings) in a readable digital format any documents; keep a digital record of the proceedings; automatically determine the specialization of judges by category of cases and distribute cases among the judges of the relevant court; administer the issuance of digital writs of execution and subsequent monitoring of their legal fate. Certain functions of the judicial-AI can be performed in voice assistant mode (like Siri, Alice or Sberbank Online).

The next stage will be the participation of AI in the legal assessment of a number of evidence in the case, in particular, to determine the category and legal properties of the transaction (for example, a major transaction or a transaction with an interest in a corporate dispute, the form of the transaction, the date of the transaction and the authenticity of the electronic signature); check the calculation of claims (including the amount of contractual penalty, real damage or loss of profit); determine the statute of limitations and the period for filing a claim; offer to reconcile the parties (such as the options for settlement options or legal perspectives on the use of mediation procedures); and calculate deepfake and other forgeries.

Artificial intelligence at this stage is possible only in liaison with a human judge—by analogy with a co-robot (a robot controlled by a human). It is a combination of AI working either in tandem with a human judge or under judge's control in the field of legal-machine processing and evaluation of evidence as information about the facts on which the parties justify their position in court.

For the gradual introduction of judicial-AI, the following is needed:

- digitisation of all textual documents coming into court into the electronic machine-readable form, respecting uniform formats and standards, with the possibility of their subsequent transformation into other formats (namely, PDF, RTF, DO and DOCX);
- eliminating the digital divide among civil actors and ensuring technological access to digital justice;
- providing AI access to the integration bus of the IEIS (inter-agency electronic interaction system) and to all kinds of Big Data;
- developing a cloud-based model of AI working remotely through the Internet telecommunication network and administering it by the highest courts, the Supreme Court and the Constitutional Court, respectively;
- adoption of regulatory legal acts regulating the possibility of transferring part of the functions of a human judge to artificial intelligence;
- formation of a unified space of trust in the digital environment and a legal culture of application of artificial intelligence.

Recognition of the possibility of artificial intelligence work in court entails legal-machine processing and evaluation of evidence in a case (information about facts), based on which a court decides.

As a general conclusion, it follows that technology of judicial artificial intelligence work should be open, reliable and transparent for all citizens, economic agents and society. This approach will ensure public confidence in the court and in the modern information technology that is being introduced into its work, such as artificial intelligence and cloud computing.

13 Conclusion

As a result of the introduction of information technology into the activities of Russian courts, the following procedures has been optimized: electronic interaction between participants in the process and the court, document management in court, storage of court files, and so on. Internet portals of courts of all levels increased accessibility to information on court activities. In addition to traditional elements of e-justice, the latest digital technologies, cloud computing (storage and use of electronic), online (remote) courts and artificial intelligence (companion judge-human) are being introduced. The approaches to the legal assessment of electronic evidence (electronic documents, messages in social networks and business messengers, web-archives) and court hearings are changing, some of which take place without summoning the parties and without face-to-face court hearings. The presented analysis defines the possible ways of development of modern electronic justice in Russia. The suggested approaches and successful examples of Russian electronic justice can be used by judicial systems of other states.

References

Alexandrov A (2020) "Trials will be connected to personal computers. Supreme Court of Russia adapts work to antiviral restrictions." December 8, 2020. https://www.kommersant.ru/doc/4604397

Brandon J (2018) "Terrifying high-tech porn: Creepy 'deepfake' videos are on the rise (English)." Fox News, February 16, 2018. https://www.foxnews.com/tech/terrifying-high-tech-porn-creepy-deepfake-videos-are-on-the-rise

Deshko I, Kryazhenkov K, Rosenberg I, Tsvetkov V (2016) Cloud platforms and services. MAKS Press, Moscow

Kashanin AV, Kozyreva AB, Kurnosova NA, Malov DV (2020) In: Kashanin AV (ed) Information technology in justice: status and prospects. Russia and the world. Analytical report, Moscow. http://xn%2D%2Do1abds.xn%2D%2Dp1ai/wp-content/uploads/2020/07/w-informacionnie-tehnologii-v-pravosudii.pdf

Laptev VA (2017) Electronic evidence in arbitration proceedings. Russian Justice 2:56–59

Laptev VA (2021) "Artificial intelligence in court: how it will work." Pravo.ru, June 2, 2021. https://pravo.ru/opinion/232129/

Laptev VA, Solovyanenko NI (2017) Electronic justice and electronic document management as a condition for modernization of the regulatory environment for business. Russian Judge 2:16–21

Laptev VA, Solovyanenko NI (2019a) In: Chucha S (ed) Electronic justice. Electronic document management: a scientific and practical manual. Prospect, Moscow

Laptev VA, Solovyanenko NI (2019b) "court cloud": legal issues of structuring and data protection. Actual Prob Russian Law 6:195–204. https://doi.org/10.17803/1994-1471.2019.103.6.195-204

Momotov V (2020) "Speech by V.V. Momotov, Chairman of the Council of Judges of the Russian Federation. Momotov at plenary session on "Prospects for use of artificial intellect in the judicial system of the Russian Federation", Qatar." Council of Judges of the Russian Federation, February 26, 2020. http://www.ssrf.ru/news/lienta-novostiei/36912

Obidin KV (2020) Electronic evidence: a necessary stage in the development of criminal proceedings. Actual Prob Russian law 11:198–206

RAPSI (2015) "Russian courts will switch to e-justice system in 2016." September 7, 2015. http://rapsinews.ru/judicial_news/20150907/274526705.html

Reese G (2009) Cloud application architectures. Wiley, Hoboken

Romanenkova SV (2013) Concept of electronic justice, its genesis and implementation in law-enforcement practice of foreign countries. Arbitr Civil Procedure 4:26–31

Sharayev S (21 July 2010) "E-justice: results and prospects. Interview." Garant.ru. http://www.garant.ru/company/cooperation/gov/action/regional/271319/

Sharifullin RA, Burganov RS, Bikmiev RG (2018) Elements of electronic justice. Russian Judge 6: 57–62

Sozankova E (2019) "Dipfakes: how neural networks copy reality" Screenlifer. Accessed May 21, 2021 https://screenlifer.com/trends/dipfejki-kak-nejroseti-kopiruyut-realnost/

Supreme Court of the Russian Federation (2021) *Concept of Informatization of the Supreme Court of the Russian Federation*. http://www.vsrf.ru/files/29675/

Supreme Court of the Russian Federation (n.d.) Accessed June 1, 2021. http://www.cdep.ru/index.php?id=79

Voronin MI (2019) Electronic evidence in the CPC: to be or not to be? Lex Russica 7:74–84

Zhurkina OV (2020) Electronic evidence: the concept and signs. Russian Justice 9:44–46

Zscheige H (6 February 2020) "How chatbots save lawyers time and money." Pravo.ru. https://pravo.ru/news/218282/

MIX
Papier aus verantwortungsvollen Quellen
Paper from responsible sources
FSC® C105338

If you have any concerns about our products,
you can contact us on
ProductSafety@springernature.com

In case Publisher is established outside the EU,
the EU authorized representative is:
Springer Nature Customer Service Center GmbH
Europaplatz 3, 69115 Heidelberg, Germany

Printed by Libri Plureos GmbH
in Hamburg, Germany